THE LIVING AFRICA SERIES

Book 4

GERD+

AFRICA'S FUTURE ENERGY HUB

Lakew Gebrekirstos Buli

2025

Published by **Lakew Gebrekirstos Buli**
Addis Ababa, Ethiopia
Printed in Ethiopia

ISBN: 978-99990-56-84-7
First Edition, 2025
Registered with the Ethiopian Intellectual Property Authority (EIPA)

The Living Africa Series

Strategic Africa • Leadership for a Living Continent • Managing Africa's Project Portfolio • GERD+ • The Pan-African Canal • Circular Africa • Skyscraper Africa • Electrify Africa • The Vision • The African Living System

Note:

The ideas and concepts expressed in this book are original works of the author, written to inspire Africa's self-reliance, unity, and transformation.

For permissions, partnerships, or distribution inquiries, contact:
lakewg@gmail.com

Table of Contents

PREFACE

There are moments in history when a single project can change how a continent sees itself. The Grand Ethiopian Renaissance Dam (GERD) is one of those moments. But GERD+ is more than a dam. It is a new way of thinking about power, energy, and possibility.

GERD+ is the author's original concept. It is not a government plan, nor an engineering design from any institution. It is an idea to show what Africa can do when it connects its own rivers, its own sunlight, and its own intelligence. It is a symbol of a living Africa - creative, united, and self-reliant.

At its heart, GERD+ transforms Ethiopia's hydropower into a continental system of clean energy. It connects floating solar panels, hydrogen production, and ammonia synthesis into one living network. It links to the Pan-African Canal and carries energy, food, and water across borders. It turns Africa's geography into its strength.

This book is not only about technology. It is about the rebirth of confidence. The story of GERD+ is the story of how Africa can use its own natural rhythm to power its future. Hydropower and solar energy together can give us the electricity to produce hydrogen, ammonia, fertilizers, and fuels for clean industries. But beyond that, they can power our minds - the most important energy of all.

Africa has everything it needs to lead a new energy civilization: sunlight, water, land, and young people full of ideas. What we need now is integration - not imitation. We must connect what we already have, value what we already know, and build what we truly need.

The Pan-African Canal, the hydrogen corridors, the green cities, and the circular systems discussed here are also the author's creations - not to describe what is, but to inspire what can be. They are sketches of a possible future, a living proof that imagination can be our greatest infrastructure.

Energy is more than electricity. It is motion, growth, and renewal. When used wisely, it can create peace, feed nations, and give dignity to work. When owned collectively, it can unite people across borders.

This book belongs to every young African who believes that our rivers can light our homes, that our deserts can store our power, and that our unity can power our future. GERD+ is not a dream - it is a direction. A living idea for a living continent.

INTRODUCTION

This book is not about a government project or a political agenda. It is about an **idea** - an idea born from a vision of how Africa can begin to **live again**. The concepts of **GERD+** and the **Pan-African Canal** are not institutions or policies; they are **symbols of possibility**, created by the author to demonstrate how a living continent can transform itself using its own wisdom, water, and light.

GERD+ stands for more than a dam. It represents a new relationship between energy and life - where power does not destroy, but restores; where technology does not exploit, but extends nature's intelligence. The "plus" is not a technical addition; it is a moral one. It means Africa's energy systems can go beyond production to regeneration, beyond survival to renewal.

The Pan-African Canal, too, is more than a channel of water. It is a **vision of connection** - a continental artery that links coasts, climates, and communities. It shows how geography can unite rather than divide, how shared water can become shared destiny. Together, GERD+ and the Canal form a living model of what Africa could become when it begins to design from its own rivers, skies, and spirit.

This book is therefore not a plan to be implemented but a **philosophy to be understood** - a framework through which Africa can imagine itself differently. It challenges the old narrative that progress must be imported, that modernity must come from elsewhere. Instead, it offers a quiet confidence: that Africa already holds within its landscapes and cultures the blueprints for a sustainable civilization.

In these pages, engineering becomes poetry, and light becomes meaning. Hydropower, hydrogen, and forests are not treated as separate topics, but as parts of one circle - a living system where energy, ecology, and ethics coexist.

The purpose of this book is to inspire reflection, not instruction; imagination, not imitation. It invites readers - African and global alike - to see the continent not as a place waiting to be developed, but as a **living intelligence** waiting to be remembered.

Africa's renaissance will not come from wealth alone, but from wisdom - from systems designed with care, rooted in nature, and guided by spirit. This is what GERD+ and the Pan-African Canal stand for: not monuments of ambition, but **mirrors of balance**, showing how life can once again flow freely across a continent that was always meant to lead through light.

Author's Note

I write as a systems thinker, not as a specialist. My purpose is to connect existing knowledge, not to replace experts. Africa already has immense scientific and technical wisdom; what we often lack is a framework that links them together. These books are my attempt to draw those links - to help others see the system, not just the parts.

1.0 THE POWER WITHIN AFRICA

Africa is often described as a continent of potential. But potential means little until it becomes power. Power in its truest sense is not only electricity - it is the ability to act, to decide, and to create. Africa's transformation begins the moment it believes that its power is already within.

The rivers that cut through our highlands, the sunlight that touches our deserts, and the winds that pass through our valleys are not problems to manage; they are gifts to organize. Energy is not something Africa must import. It is something Africa must harmonize.

The story of Africa's energy is a story of contrasts. Across vast lands, millions live in darkness while vast natural light remains unused. The continent that carries one-fifth of the world's population generates less electricity than some single countries outside it. But this imbalance is not a curse; it is a call. A continent that has lived without full power for so long understands better than anyone the value of light.

Africa's future will not come from copying what others have done. It will come from reimagining what energy means in the African context. Energy is not just megawatts and grids - it is water flowing through life, sunlight stored in food, and movement turned into creation. The African definition of energy must be holistic, human, and circular.

The Grand Ethiopian Renaissance Dam (GERD) represents a new chapter in this story. It is not only a hydropower project; it is a statement that Africa can build something that lasts, something large, something its children can inherit with pride. But even GERD itself is not the destination. It is a beginning. That is why the author created **GERD+**, to demonstrate how one dam can become a continental energy system, feeding not only Ethiopia but Africa's shared vision of clean, connected growth.

The "+" in GERD+ symbolizes addition - the addition of sunlight, hydrogen, knowledge, and cooperation. It stands for expansion beyond boundaries, beyond national projects into continental purpose. It reminds us that Africa's power multiplies when its nations act together.

In the 21st century, energy defines independence. Those who control their energy control their destiny. Africa must no longer remain a passive consumer of energy systems designed elsewhere. It must become a designer of its own. The continent's strength lies in integration: linking hydropower with solar, connecting river basins to ports, merging electricity with food and water security.

Power is not only in the turbines; it is in the people. Africa's youth are the real grid - vibrant, intelligent, and full of current. When we connect their creativity to our natural resources, the result will be a living, breathing energy network powered by purpose.

The Power Within Africa is already here. It flows through our rivers, our forests, our cities, and our dreams. GERD+ is only a reflection of what already exists - a reminder that the light we are looking for has always been ours.

1.1 The Sleeping Giant: Africa's Untapped Energy

Africa is a giant that has slept through centuries of abundance. Beneath its soil lies sunlight waiting to be turned into electricity. Across its rivers flows energy strong enough to light an entire continent. Yet the giant still rests - watching others build from the resources it gave away.

For too long, Africa's wealth has been measured by what leaves its borders. Minerals, oil, gas, and raw materials have been extracted, shipped, and sold to power other civilizations. What remains behind are the shadows of dependence - bright lights elsewhere, dim hopes here. But the story does not end there. The giant is stirring.

To awaken Africa's power, we must first redefine what energy means to us. True energy is not just the fuel that turns turbines or the current that lights homes. It is the living force that transforms a nation from survival to dignity. It is the power to feed, to teach, to move, and to create.

Africa's rivers alone could produce more than 400,000 megawatts of clean power - enough to supply most of the continent's demand. Its deserts receive more solar radiation than almost any other place on Earth. The winds along its coasts, the geothermal heat beneath its rift valleys, and the biomass from its farmlands all speak the same message: *there is no energy shortage here - only a shortage of integration and vision.*

The sleeping giant does not lack strength; it lacks coordination. Each country plans in isolation, each region acts alone, and so the power remains fragmented. Our challenge is not production - it is connection.

GERD+ is a wake-up call to this sleeping giant. It is a model of what can happen when Africa chooses cooperation over competition. By combining hydropower, floating solar, and hydrogen, GERD+ shows that we can multiply value rather than divide it. It proves that when we design as a continent, not as individual nations, the sum becomes far greater than its parts.

Electricity is not the only thing that flows through our rivers. Ideas do too. The more we share them, the more powerful we become. Imagine a future where Ethiopia's hydropower connects to Sudan's farmlands, where Kenya's engineers work alongside Ghana's technicians, where North Africa's solar fields power Southern Africa's industries. That is not a dream - it is the logic of unity.

The sleeping giant is awakening. Across Africa, young innovators are already building solar microgrids, recycling waste into energy, and designing clean transport systems. Every small initiative is a spark; together they can ignite a continental transformation.

When Africa rises, it will not rise through extraction, but through creation. The continent's energy will not come from digging deeper into the ground, but from thinking deeper about the sky, the water, and the life that surrounds us.

The giant is not waiting for permission. It is stretching, breathing, and remembering its strength. And once fully awake, the world will see that Africa was never poor - it was only preparing to rise.

1.2 The Meaning of Renaissance in Energy

The word *Renaissance* means rebirth. It speaks of light returning after darkness, of minds reopening to knowledge once forgotten, and of people rediscovering their power to shape destiny. In Europe, the Renaissance was about art and science. In Africa, our renaissance must begin with energy - because energy is the foundation of every other awakening.

Without light, there is no learning. Without power, there is no production. Without clean energy, there is no lasting prosperity. To build a modern, living Africa, we must first light it - not only with electricity, but with the energy of imagination and unity.

A true African energy renaissance is not about copying the grids and systems of others. It is about creating energy that reflects who we are and how we live. It is about designing power systems that serve our people and our planet, not markets alone.

For centuries, Africa has exported its raw energy - in the form of oil, gas, and coal - to fuel the growth of others. In return, it imported finished goods, technology, and dependency. This cycle must end. The African Renaissance in energy is a declaration that we will no longer export our strength in raw form. We will refine it, transform it, and use it to build a continent of our own design.

GERD+ represents that rebirth. It is a living idea that combines Africa's greatest gifts - water, sunlight, and intelligence - into one integrated system. It shows that progress does not require pollution. It proves that development can flow in harmony with nature, not against it.

In this renaissance, water is not just for irrigation; it is for transformation. The rivers of Africa are no longer boundaries; they are bridges of energy and trade. The deserts are no longer empty; they are vast power fields of sunlight. The winds are no longer invisible; they are messengers of motion.

But the true source of this rebirth lies not in technology - it lies in consciousness. A continent rises when its people believe they can rise. The African Renaissance in energy begins in the mind before it reaches the grid. It starts when a young engineer decides to innovate rather than imitate. It grows when a government invests in collective good instead of private gain. It flourishes when communities understand that energy is not a product but a shared right and responsibility.

Renaissance means renewal - of purpose, of dignity, and of belief. It means reconnecting with our past wisdom and combining it with future intelligence. Africa's ancient civilizations once mastered the balance between water, agriculture, and architecture. GERD+ is a continuation of that heritage - using modern tools to express an old truth: that life itself is energy in motion.

The African energy renaissance will not be loud; it will be steady. It will not depend on slogans; it will depend on systems. It will grow quietly in every solar panel installed, in every hydrogen plant built, in every mind that chooses collaboration over division.

The renaissance has already begun. Each sunrise over the Nile, the Congo, and the Niger reminds us that the power we seek has always been ours - we only needed to believe it again.

1.3 From Dependency to Self-Reliance

For generations, Africa has lived in the shadow of dependence. Our lands are rich, yet our systems depend on aid. Our rivers are full, yet our cities wait for power. Our people are capable, yet our industries rely on imported technology. This dependence is not natural; it was built. And what is built can be rebuilt.

Dependency begins with doubt. When a nation doubts its own capacity, it looks outward for strength instead of inward. It imports not just goods, but confidence. It learns to wait for solutions instead of designing them. And slowly, the habit of waiting becomes part of its culture. But Africa is reaching a turning point - the moment when waiting is no longer an option.

Self-reliance begins the moment we trust ourselves. It is not isolation; it is empowerment. It means producing what we can, learning what we must, and trading what we choose. It means seeing cooperation as a partnership, not as charity.

Africa's path to energy independence begins with understanding that the continent already holds everything required for transformation. The sun above us, the water beneath us, the minds among us - these are the raw materials of the future. We do not need to wait for others to deliver what nature has already given us in abundance.

GERD+ is a demonstration of that self-reliance. It takes what Ethiopia already has - water, terrain, sunlight - and transforms it into a continental energy model. It proves that Africa can innovate on its own terms, using its own resources to solve global challenges. The integration of hydropower, floating solar, and hydrogen production within GERD+ is not only a technical innovation; it is a statement of independence. It says that Africa can create technology that belongs to Africa.

Dependency is expensive. Every imported kilowatt, every foreign-designed power plant, every borrowed technology carries a cost far beyond money - the cost of control. When others own the systems that power your economy, they also shape your choices. True independence begins when we design, build, and maintain our own systems of energy.

Self-reliance also requires unity. A single nation may achieve energy growth, but only a united continent can achieve energy security. The Pan-African Canal and the GERD+ hydrogen network are not just projects; they are frameworks of cooperation. When energy flows across borders, peace follows. When nations share power, they also share stability.

To be self-reliant is not to reject partnership; it is to define it. Africa can and must collaborate with the world, but from a position of strength - as a continent that contributes, not one that depends. The partnership Africa seeks is one of equality: technology for access, investment for value, and respect for sovereignty.

The age of dependency is closing. The age of design is beginning. Africa is no longer a recipient of energy ideas; it is becoming their author. GERD+ is not only an energy project - it is a symbol of what happens when Africa believes in its own capacity.

When we own our power, we own our future. Self-reliance is not an act of separation; it is the beginning of true connection - connection to our resources, to our people, and to the purpose that unites us all.

1.4 The Language of Water and Light

Water and light are the oldest languages on Earth. Long before humans built cities or wrote books, rivers spoke through their flow, and sunlight taught life how to grow. In Africa, these two elements have always been more than resources - they are teachers, healers, and messengers of balance.

To understand Africa's energy future, we must first learn to listen again to this language. Water tells us how to move. It never forces its way; it finds its path. It shapes mountains over centuries without noise or anger. It connects everything it touches - soil, plants, people, and oceans - reminding us that all life is part of one continuous flow.

Light, on the other hand, teaches us how to renew. Every morning it begins again, no matter how dark the night was. It gives without expecting, and in its quiet generosity it powers every seed, every cell, and every thought. Together, water and light are nature's perfect partnership - one flowing, one shining - both creating.

Africa's energy transformation must follow their example. Hydropower and solar are not just technologies; they are reflections of life itself. When we combine them wisely, we are not merely generating electricity - we are aligning with nature's rhythm. That is the philosophy behind **GERD+**. It unites the steady strength of water with the endless abundance of sunlight. It turns natural harmony into practical design.

GERD+ speaks this language fluently. The dam stores the motion of water; the floating solar panels capture the silence of light. Together they produce clean power for hydrogen, ammonia, and industry - all flowing in a circular system. Nothing is wasted; everything is reused. This is how a living continent should function: circular, regenerative, intelligent.

In this system, water does more than spin turbines - it cools the land, nourishes farms, and stabilizes climate. Light does more than charge panels - it feeds cities, warms homes, and powers dreams. The combination of both gives Africa an energy foundation that no fossil fuel can match.

But the language of water and light is not only physical; it is spiritual. It reminds us of humility and patience. To work with nature is to listen to it, not to command it. Africa's ancient civilizations understood this truth. They built their lives around the cycles of rain and the journey of the sun. GERD+ revives that wisdom through modern tools - hydropower plants, solar fields, and hydrogen systems that speak the same ancient language in a new form.

When we build with respect for the natural order, our systems become more than machines - they become ecosystems. Every dam, every canal, every solar field must live in balance with its surroundings. The future of African energy is not just about expansion; it is about harmony.

Water and light are already guiding us. They do not need translation; they need attention. If we follow their rhythm - slow, strong, and steady - Africa will never lack power again. The rivers will teach us how to connect, and the sunlight will remind us that renewal is always possible.

Africa's energy renaissance is not about domination of nature; it is about communion with it. Water and light have always been on our side. Now it is our turn to speak their language again - through design, through unity, and through purpose.

1.5 The Grand Ethiopian Renaissance Dam: A Symbol of Collective Will

The Grand Ethiopian Renaissance Dam stands as one of the most powerful symbols of Africa's new era. Rising from the highlands of Ethiopia, it is not merely a wall of concrete or a collection of turbines - it is a monument to determination, unity, and belief. It represents what happens when a nation decides to write its own story, using its own hands, with its own vision.

The GERD was built without dependency, financed by the Ethiopian people themselves. Farmers, workers, and professionals all contributed what they could. Some gave money, others gave time, others gave courage.

That collective spirit is what gives the dam its true strength. It is not powered by water alone - it is powered by faith.

For centuries, the flow of the Nile symbolized Africa's dependence on forces beyond its control. GERD changed that narrative. It showed that Africa can command its own destiny, that it can manage its own resources responsibly, and that it can generate not only power, but pride.

Yet GERD is not the end of the story. It is the foundation of a larger vision - **GERD+** - an idea that expands the meaning of the dam beyond electricity. In the GERD+ concept, the dam becomes a living system: it produces hydropower, supports floating solar platforms, enables hydrogen production, and fuels regional cooperation. It is both a local achievement and a continental seed.

The + in GERD+ means addition - not just in power output, but in imagination. It adds innovation to tradition, light to water, and unity to purpose. It connects energy to food, electricity to industry, and nations to one another. It represents a living bridge between nature and technology, showing that Africa's development can grow from its own rhythm.

The true greatness of the dam is not its size, but its symbolism. It proves that collective will can overcome any limitation. When people believe in a shared vision, they can move mountains, carve rivers, and light continents. GERD tells every African child that our dreams do not have to be imported - they can be built right here, on our own land, by our own minds.

GERD also reminds us that unity is strength, and cooperation is survival. The energy it produces can illuminate homes across borders. Its lessons can inspire every nation that seeks transformation through self-reliance. It teaches that the greatest resource we possess is not water or sunlight - it is willpower.

The Grand Ethiopian Renaissance Dam is not only Ethiopia's story. It is Africa's story of awakening. It stands as a mirror reflecting what the continent can achieve when it refuses to be divided by fear or hesitation. It marks the beginning of a new kind of independence - one built on creation, not conflict; on innovation, not imitation.

When future generations look at the GERD, they will see more than an engineering marvel. They will see a message written in stone and water: that when Africa unites its purpose with its power, nothing is impossible.

2.0 THE BIRTH OF GERD+: A VISION BEYOND A DAM

Every great transformation begins with a question. The question that gave birth to **GERD+** was simple: *What if a dam could do more than produce electricity?*

That question turned a structure into a system, and an engineering project into a living idea. GERD+ was born not as a physical expansion, but as a conceptual one - a model that reimagines how Africa can turn one achievement into a thousand possibilities.

The Grand Ethiopian Renaissance Dam was a historic milestone, but GERD+ looks beyond the turbines. It asks: *Can water power also create hydrogen? Can a reservoir host floating solar fields? Can energy become the backbone of continental trade, irrigation, and industry?* GERD+ answers yes to them all.

The idea is to use what already exists - water, sunlight, and intelligence - to build a complete, circular energy ecosystem. Hydropower becomes the base, solar energy becomes the extension, and hydrogen becomes the bridge that connects electricity to transport, fertilizer, and industry. Together, they form a living network of renewable energy.

The "+" represents connection. It is the addition of new dimensions - technology, cooperation, and imagination. It transforms GERD from a national dam into a continental model. It links Ethiopia's energy heart to the Pan-African Canal, to hydrogen export corridors, and to industrial hubs powered by clean ammonia and green electricity.

But GERD+ is not only about systems; it is about symbolism. It shows that Africa can design its own solutions, instead of waiting for others to design them. It reflects confidence - the confidence that Africa can lead the next global energy era not by competing, but by rethinking.

This idea was born from the author's conviction that every African project should be a seed, not an end. GERD+ is such a seed - an open concept designed to grow across borders, inspiring new hydropower expansions, floating solar fields, hydrogen plants, and irrigation corridors throughout the continent.

The birth of GERD+ also marks a philosophical shift. It changes how we see dams: not as barriers across rivers, but as platforms of life. It changes how we see energy: not as a commodity to sell, but as a force to build civilizations. And it changes how we see Africa: not as a consumer of ideas, but as their creator.

GERD+ is Africa's invitation to the future - a future powered by its own light, guided by its own rivers, and built by its own people. It is proof that when imagination meets engineering, a continent can rise beyond its challenges and shine for the world to see.

2.1 Why "GERD+"? – From Power to Hydrogen

The idea of **GERD+** was born from a realization: that power alone is not enough. Electricity lights homes, but hydrogen fuels industries, transportation, and global trade. To transform a continent, Africa must move from producing electricity to producing energy in all its living forms.

Hydropower is strength, but hydrogen is strategy. Hydropower provides clean electricity, while hydrogen stores that electricity, transports it, and multiplies its value. When we combine them, energy becomes not just a service - it becomes a system of independence.

The Grand Ethiopian Renaissance Dam already produces thousands of megawatts of clean energy. But electricity, by nature, cannot travel far or be stored easily. Hydrogen changes that. It turns surplus hydropower into a portable, storable, and tradable energy source. It connects rivers to ports, farms to factories, and nations to markets.

In **GERD+**, hydropower becomes the mother source - the origin of green hydrogen through electrolysis. Solar panels floating on the reservoir strengthen it by adding daytime generation while reducing water evaporation. Together, they produce electricity that splits water molecules - hydrogen for fuel and ammonia for fertilizer. This cycle links power to food, energy to agriculture, and production to prosperity.

Hydrogen is not a foreign technology. It is a simple element, the most abundant in the universe - the same that burns in the stars above Africa. To create it, we only need what Africa already has: water, sunlight, and vision. The combination of hydropower and solar energy makes GERD+ one of the cleanest and most sustainable hydrogen models ever imagined.

But the "+" in GERD+ is not only scientific; it is symbolic. It represents *addition* - adding value, knowledge, and purpose. It stands for Africa's capacity to expand ideas, not just replicate them. GERD+ proves that a dam can be more than a generator; it can be a platform for continental transformation.

Through hydrogen, Africa can fuel its own industries, replace imported fossil fuels, and export clean energy to the world. It can power buses, trains, and ships with its own molecules of light. It can turn energy from something consumed into something created.

In the global shift to a green economy, hydrogen is becoming what oil once was - the foundation of industrial and geopolitical power. GERD+ positions Africa at the center of this new energy map. It transforms Ethiopia from a regional producer into a continental and global energy hub, capable of exporting clean ammonia and hydrogen through the Pan-African Canal and the Red Sea ports.

This is why GERD+ matters. It is not about building more infrastructure; it is about building more intelligence. It is the step from using power to creating possibility - from electricity that flows for hours to energy that lasts for generations.

Hydrogen is the link between Africa's natural abundance and its industrial future. GERD+ gives that link a name, a form, and a direction - proving that the continent's energy renaissance begins not in scarcity, but in imagination.

2.2 Transforming Hydropower into Continental Energy

Hydropower is one of nature's most generous gifts to Africa. It flows quietly through the continent's veins - in the Nile, the Congo, the Niger, the Zambezi, and hundreds of smaller rivers that breathe life into valleys and plains. Yet for too long, this gift has been used only in fragments, serving nations but not the continent. GERD+ transforms hydropower from a local source into a continental system.

When water falls through turbines, it releases not just electricity - it releases opportunity. The Grand Ethiopian Renaissance Dam captures that opportunity at scale, but GERD+ extends it further. It links hydropower to hydrogen production, to solar energy, and to Africa's interconnection networks, turning each drop of water into part of a continental flow of energy.

Hydropower's greatest strength is its stability. Unlike the sun, it runs day and night. It balances the grid, stores potential energy, and regulates supply. But its true power comes when combined with other renewables. GERD+ creates that balance. When sunlight fades, water continues. When rainfall slows, floating solar arrays step in. Together, they form a self-correcting system - flexible, renewable, and circular.

This transformation is not about building more dams, but about thinking differently about the ones we have. Across Africa, hydropower reservoirs can become hybrid energy parks - each integrating floating solar panels, electrolysis systems, and green ammonia plants. In this way, Africa's rivers can power not only local grids but regional industries, export markets, and even global clean fuel corridors.

The dam becomes the heart of a living organism. The turbines generate electricity for domestic use and hydrogen production. The hydrogen, combined with nitrogen from the air, becomes ammonia - a transportable form of clean energy and fertilizer. The ammonia feeds agriculture, while the revenue feeds development. Every cycle adds value. Nothing is wasted.

The beauty of hydropower lies in its harmony with Africa's geography. From the Ethiopian highlands to the Great Lakes, from the Congo Basin to the Orange River, the potential is immense. If coordinated through a continental framework - a network of smart grids, hydrogen pipelines, and canal systems - Africa could power itself many times over.

GERD+ envisions that coordination. It connects hydropower stations across countries, integrates them with solar corridors, and channels surplus energy into hydrogen hubs. It turns scattered generation into a united system - the African Energy Belt.

Transforming hydropower into continental energy is not a technical challenge; it is a strategic decision. It requires seeing Africa not as 54 separate grids, but as one living network powered by rivers that have never recognized borders. It requires political will, shared standards, and visionary leadership - the kind that thinks in centuries, not terms.

This transformation also brings a moral dimension. Hydropower is clean, renewable, and regenerative. It replaces fossil fuels without damaging ecosystems. It respects the rhythm of nature while empowering humanity. Through GERD+, hydropower becomes more than an energy source - it becomes a philosophy of balance, creation, and continuity.

When Africa learns to flow like its rivers, nothing will stop its progress. The continent's future energy map will not be drawn by lines of pipelines or cables alone, but by the currents of cooperation that link every drop of water to every spark of light.

2.3 The Role of Floating Solar Systems

Africa's sunlight is its most abundant inheritance. Every day, the continent receives enough solar energy to power its entire population many times over. Yet much of that light goes unused, reflected off water and land, waiting to be captured. The **floating solar systems** of GERD+ are designed to turn that waiting light into life.

Floating solar panels - sometimes called "floatovoltaics" - are solar arrays installed directly on water surfaces such as reservoirs, lakes, or canals. They do more than generate electricity. They create synergy. When paired with hydropower dams, they form one of the most efficient and sustainable energy systems known to humanity.

In the GERD+ concept, the reservoir behind the Grand Ethiopian Renaissance Dam becomes a dual-purpose engine. The dam harnesses the movement of water; the floating panels capture the energy of light. Together, they expand the plant's power output while protecting the environment. Covering just 15 to 20 percent of the reservoir's surface with solar panels can generate several thousand additional megawatts - without building new dams or flooding new land.

But the benefits go beyond electricity. Floating panels reduce water evaporation by shading the surface, conserving millions of cubic meters of freshwater each year. That water, in turn, supports irrigation, ecosystems, and downstream communities. The panels also keep the water cooler, improving the dam's efficiency and extending the life of its turbines. Nature and technology, once again, work together.

Floating solar is also modular and scalable. It can be installed quickly, expanded gradually, and integrated seamlessly with existing infrastructure. In the GERD+ system, floating solar fields are connected directly to electrolysis units for hydrogen production. During the day, sunlight powers electrolysis; at night, hydropower takes over. The combination ensures constant, 24-hour clean energy production - a key foundation for hydrogen and ammonia synthesis.

This model can be replicated across Africa. Many existing hydropower reservoirs - from Uganda to Zambia, from Ghana to Angola - can host floating solar systems. These hybrid installations would transform static water bodies into dynamic energy platforms, forming a continental network of green electricity. Each site becomes both a storage lake and a power station, feeding into regional grids and hydrogen pipelines.

The floating solar approach also minimizes land use conflicts. In many countries, vast tracts of land are needed for agriculture, forestry, and settlement. By using water surfaces instead, Africa can expand its renewable energy capacity without competing for fertile land. This aligns perfectly with the continent's vision of harmony between food, energy, and environment.

More than a technology, floating solar represents a philosophy - the belief that progress does not have to harm nature. It is quiet, graceful, and adaptable. It reflects the sunlight Africa has always carried but never fully claimed.

When we look at a calm reservoir covered with panels, we are seeing more than reflection. We are seeing transformation. Each panel becomes a mirror of Africa's awakening - a reminder that light can be both gentle and powerful, and that innovation, when aligned with nature, can heal as it builds.

GERD+ is the beginning of that alignment. Through floating solar systems, water and light unite once again to power a new African era - one where energy is endless, balanced, and beautifully African.

2.4 The Integration of Hydrogen, Ammonia, and Hydropower

In nature, everything is connected. Rain becomes river. River becomes energy. Energy becomes life. The **GERD+** system follows this same rhythm - combining water, sunlight, and science to create a continuous circle of clean power. At its core lies the integration of **hydropower, hydrogen, and ammonia** - three forces that, when united, can fuel an entire continent.

Hydropower provides the foundation. It delivers steady, renewable electricity every hour of the day. This electricity is the heartbeat of GERD+, supplying both homes and industries while feeding the process that produces hydrogen through **electrolysis** - the simple act of splitting water into hydrogen and oxygen using clean power.

The hydrogen created from this process becomes the continent's new energy currency. It can be stored, transported, or converted into other forms of energy. But its greatest strength lies in combination - when merged with nitrogen from the air, hydrogen becomes **ammonia**. This single transformation unlocks two revolutions: one in energy, the other in agriculture.

Ammonia is easier to transport than hydrogen, and it already has a global market as a fertilizer. In the GERD+ framework, ammonia becomes both **fuel and food** - powering ships and industries while feeding Africa's soils. A single molecule of hydrogen, therefore, serves multiple purposes: it powers vehicles, strengthens crops, and creates value chains that connect farmers, engineers, and exporters.

Hydropower, hydrogen, and ammonia form a cycle of renewal. The dam provides electricity for electrolysis. Electrolysis produces hydrogen. Hydrogen combines with nitrogen to form ammonia. Ammonia fuels power plants, fertilizes fields, and generates income. The revenues then support further energy expansion, creating a self-sustaining circle of progress.

This integration also stabilizes Africa's energy system. Hydropower ensures base load capacity; hydrogen stores surplus electricity for future use; and ammonia acts as a carrier for export and domestic supply. The result is an energy network that can adapt to seasons, market shifts, and regional demand without depending on fossil fuels or external systems.

The GERD+ vision places Ethiopia - and Africa - at the forefront of a new industrial model: **the green molecule economy**. It is not an imitation of the carbon age; it is its replacement. Here, water is the new oil, sunlight the new refinery, and collaboration the new currency.

By combining these elements, Africa gains more than energy. It gains sovereignty. The continent can produce its own fertilizers, powering food security without relying on imported chemicals. It can export clean ammonia to global markets as a fuel for ships and industries. It can store its energy in molecular form, ready to trade, use, or reserve.

The integration of hydrogen, ammonia, and hydropower also brings harmony with nature. It creates no carbon, no smoke, no waste. The only by-products are oxygen and water - gifts that return to the atmosphere and the land. This is not extraction; it is regeneration.

GERD+ is a message that the age of burning is over. The age of transformation has begun. Africa no longer needs to dig for power; it can grow it. It can produce its own energy without harming its environment, its rivers, or its people.

Hydropower gives the rhythm. Hydrogen gives the reach. Ammonia gives the resilience. Together they create a living, breathing system - a true energy renaissance, born from Africa's own resources, guided by African minds, and built for generations yet to come.

2.5 Lessons from Global Models, Proof from Africa's Potential

Every great idea grows stronger when tested against experience. The concept of **GERD+** does not arise from imagination alone - it is inspired by what the world has already proven possible. Across continents, nations are now learning that the clean energy revolution is not a dream of the future but a practical transformation of the present.

In Mongolia, one company has built a large-scale hydrogen and ammonia complex powered entirely by renewable energy. In less than a decade, it has achieved industrial production levels once thought unreachable. This model proves that off-grid, AI-managed hydrogen systems can thrive even in remote, challenging environments. If this can be done in deserts and steppes, how much more in the fertile valleys and bright skies of Africa?

GERD+ builds upon such lessons - but with a continental vision. Unlike isolated projects, it integrates multiple renewable systems into a single, living network. Hydropower, floating solar, hydrogen storage, and green ammonia production are not separate industries; they are stages of one continuous cycle. This is what makes GERD+ uniquely African - a system that reflects the interdependence of nature, community, and innovation.

Other global experiments confirm what Africa already possesses. The world's best solar resources lie not in Europe or North America, but in Africa's deserts. Its hydropower potential is vast, with more than 400,000 megawatts still untapped. Its geological formations, such as Ethiopia's Afar salt caverns, offer ideal natural storage for hydrogen. And its youthful population brings the creativity and technical curiosity needed to turn potential into power.

Yet the greatest proof of Africa's readiness is not technical - it is moral. The continent contributes the least to global carbon emissions but suffers the most from climate impacts. This gives Africa not only the right but the responsibility to lead the clean energy era. GERD+ is not about catching up; it is about taking the lead, showing that development can be achieved without destruction.

From global models, we learn precision, management, and efficiency. From Africa's reality, we add purpose, resilience, and unity. When combined, these create a model that is not imported but reborn in African form.

GERD+ also teaches the world a different approach to energy. Most global systems were designed to serve profit; GERD+ is designed to serve life. It aligns with the rhythm of the continent - circular, renewable, and inclusive. Its aim is not to dominate markets but to empower communities, creating a chain of value that stretches from river to farm, from sunlight to home.

The proof of Africa's potential lies everywhere: in its vast water bodies, its unbroken sunlight, its human intelligence, and its growing will to cooperate. The question is no longer whether Africa can do it, but when it chooses to begin together.

The lessons from others are valuable, but the lesson from Africa is deeper: we already have what we need. When the continent unites its resources, integrates its systems, and trusts its people, the world will look to Africa not as a recipient of energy ideas but as their source.

GERD+ is that proof - a design born of Africa's rivers, refined by global knowledge, and powered by a continental dream. It is where imagination becomes industry and where the living systems of the earth meet the living systems of Africa.

3.0 HYDROGEN FOR A LIVING CONTINENT

Hydrogen is more than an element. It is the first building block of the universe, the lightest and most abundant substance in existence - the same force that powers the stars. For Africa, hydrogen represents something equally profound: the key to independence, integration, and renewal.

The continent has always carried the raw materials of life: water, sunlight, and air. Hydrogen connects them all. It is born from water, powered by light, and united with nitrogen from the air to create ammonia - the essence of food, fuel, and growth. In this sense, hydrogen is the language through which Africa can speak to the future.

Hydrogen will allow Africa to leapfrog the fossil era. It offers a clean path to industrialization - one that does not rely on imported oil or coal, and one that aligns perfectly with Africa's climate and geography. By turning renewable electricity into hydrogen, Africa can store and transport energy across thousands of kilometers, powering industries, transportation systems, and homes without harming the planet.

But hydrogen is not only a technology; it is a philosophy. It teaches flexibility, cooperation, and vision. One molecule of hydrogen can serve many purposes - to power a train, feed crops, or stabilize a power grid. In the same way, Africa's development must become multipurpose, serving both people and the planet, the economy and the environment.

The **GERD+** model transforms this philosophy into reality. It produces hydrogen using renewable electricity from hydropower and floating solar panels. It stores the hydrogen in natural salt caverns in Ethiopia's Afar region - turning geology into infrastructure. It moves it through the Pan-African Canal and pipeline network - connecting landlocked nations to global trade routes. It converts part of it into ammonia - providing fertilizer for African farmers and clean fuel for African ports.

This is how hydrogen becomes more than energy; it becomes a system of life. It connects food to power, trade to technology, and nature to civilization. It transforms energy from a national resource into a continental commons.

Africa's hydrogen revolution is not about following others; it is about defining a new path. The global north is investing in hydrogen to replace its old systems. Africa can build it as its first system - fresh, clean, and integrated. Here, hydrogen will not replace coal; it will replace poverty.

Hydrogen is also a unifier. It does not recognize borders. Its production, storage, and transport require cooperation between nations - just as rivers and trade routes do. A hydrogen-powered Africa will naturally become a connected Africa, with shared pipelines, shared industries, and shared progress.

The power of hydrogen is not in its flame; it is in its flow - the way it moves between water, air, and earth without leaving scars. It reminds us that the cleanest energy is the one that harmonizes with creation rather than consuming it.

For Africa, hydrogen is more than fuel. It is identity - the scientific expression of a spiritual truth: that we already possess within us everything we need to grow. Water, light, and knowledge. The rest is simply design.

A living continent must breathe, flow, and renew itself. Hydrogen makes that possible. It is the breath of the new Africa - invisible, powerful, and unstoppable.

3.1 Understanding Hydrogen: The Element of Life

Hydrogen is the simplest of all elements - one proton, one electron - yet it is the foundation of everything. It fuels the stars, flows through the oceans, and forms part of every living thing. It is both the beginning and the bridge of life itself.

When we look at the night sky, every glimmering star is a giant sphere of hydrogen turning into light. When we look at water, we are looking at hydrogen joined with oxygen, sustaining every plant and every breath. Hydrogen is nature's silent architect - unseen, yet present in all forms of creation.

To understand hydrogen is to understand transformation. It shows us how something small can create something vast. One molecule, when combined with others, becomes the source of power that lights the sun and shapes the clouds. Hydrogen teaches us that greatness begins with simplicity - and that energy, like life, flows through connection.

For Africa, hydrogen represents the same principle on a continental scale. It transforms what already exists - sunlight, water, and human intelligence

- into power, fuel, and opportunity. Through **electrolysis**, water (H_2O) is split into its two elements: hydrogen and oxygen. When powered by renewable electricity from hydropower and solar energy, this process becomes completely clean. No smoke, no carbon, no harm - only the unlocking of what is already within.

Hydrogen is not a fuel in the traditional sense. It is an energy carrier, a bridge between generation and use. It allows us to store the power of sunlight from day to night, to move energy from rivers to cities, from one country to another. In a continent where sunlight and water coexist in abundance, hydrogen becomes the perfect medium of balance.

But beyond science, hydrogen carries a deeper message. It reminds us that invisible forces can move the world. Like wisdom, it cannot be seen, but its effects are everywhere. It holds potential without destruction, motion without noise, and power without pollution.

The **GERD+** system embodies this principle. It uses Ethiopia's hydropower as a steady source of electricity, floating solar panels as a daytime enhancer, and electrolysis units as the translator between electricity and hydrogen. What begins as water becomes a molecule of freedom - fuel for Africa's industries, agriculture, and transport.

Hydrogen also represents fairness in nature. It can be created anywhere there is water and electricity. It belongs to no nation, no corporation, and no race. It is universal, accessible, and democratic - qualities that match Africa's vision of shared growth and equality.

To understand hydrogen is to understand possibility. It is the invisible link that can unite Africa's rivers, deserts, and coasts into one circular system of clean power. It is the element that can carry our continent's energy story into the next century.

Hydrogen is not just science - it is symbolism. It tells us that power need not be heavy, that strength can be light, and that the smallest particle can hold the largest promise.

In every drop of water lies the future of Africa. The key is to release it - wisely, sustainably, and together.

3.2 The Global Hydrogen Race

The world has entered a new energy race - not for oil, not for gas, but for hydrogen. Nations across the globe are realizing that the element once known only to scientists may hold the key to their survival in a warming planet. The hydrogen race is no longer a prediction; it is already shaping economies, industries, and alliances.

In Asia, Europe, and the Americas, governments are investing billions in what they call the "hydrogen economy." Factories are being redesigned, refineries are being converted, and transport networks are being built around hydrogen. Japan and South Korea are developing fleets of hydrogen-fueled ships. Germany and the Netherlands are constructing cross-border hydrogen pipelines. Australia is positioning itself as a global hydrogen exporter, turning its sunlight into fuel for international markets.

But while others race to adapt, Africa stands at the starting line with every natural advantage. The continent has what the world now desperately seeks: abundant solar energy, powerful rivers, open land, and access to the sea. In a global contest for clean energy, Africa begins not behind, but ahead - if it chooses to move with unity and vision.

Hydrogen is the great equalizer in this new age. Unlike fossil fuels, which are trapped beneath specific lands, hydrogen can be produced anywhere with water and electricity. This changes the geography of power. It decentralizes energy ownership and gives new opportunities to nations long excluded from industrial control. In this sense, the hydrogen race is not only economic; it is deeply political. It will reshape who leads, who follows, and who defines progress.

The **GERD+** concept places Africa - and Ethiopia in particular - at the heart of this global transformation. It connects hydropower, floating solar systems, and continental hydrogen pipelines into one living infrastructure. This model allows Africa to participate not as a supplier of raw materials but as a producer of clean energy molecules, value-added and export-ready.

The world's demand for green hydrogen and ammonia is rising rapidly. Industries that once depended on oil - shipping, fertilizers, steel, and transport - are all shifting toward hydrogen-based systems. The European Union, Japan, and many others are already searching for stable, renewable hydrogen suppliers. Africa can fill this gap, not temporarily, but permanently.

The race, however, is not only about technology; it is about timing. The nations that act now will shape the standards, control the markets, and influence the trade routes of the future. Africa must therefore run its own race - not to imitate others, but to define a distinctly African model of hydrogen development: decentralized, circular, and equitable.

The **Pan-African Canal** and its hydrogen corridors are part of that vision. By connecting inland production centers like GERD+ to coastal export terminals, the continent can create the world's first fully green trade route - powered by hydrogen, built by Africans, serving Africans and the world.

In the global hydrogen race, others compete for dominance. Africa can lead through wisdom. While others seek speed, Africa can seek balance. The continent's strength will come not from how fast it builds, but from how deeply it integrates - aligning energy with food, trade, and climate resilience.

The race for hydrogen is, in truth, a race for the future. It is not about who burns the most energy, but who understands energy best. Africa's rivers, deserts, and youth give it every advantage. All that remains is the decision to move - not separately, but together.

3.3 Why Africa Must Lead Without Mining

Africa has been called rich for centuries - rich in gold, in oil, in minerals, and in rare earths. Yet those riches have often come with a paradox: the more we dig, the poorer we become. The soil is left wounded, the people divided, and the environment stripped of its balance. The story of mining in Africa is not only a story of extraction; it is a story of lost opportunity.

The new age of energy gives Africa a chance to end that story and begin a new one. The hydrogen era does not need mining. It needs water, sunlight, and intelligence - resources that Africa holds in abundance and that can never be depleted. For the first time in modern history, Africa can lead the world without digging into its own body.

Hydrogen is not a metal. It is not a resource that must be removed from the earth. It is a living element drawn from water, powered by light. It can be produced endlessly without harm, without conflict, without loss. This is what makes it the perfect foundation for an ethical and sustainable African future.

In the past, Africa's natural wealth was exported in raw form - minerals shipped away, profits sent abroad, and communities left behind. The **GERD+** system breaks that cycle. It keeps value within the continent. It turns water into energy, energy into fertilizer, and fertilizer into food security. Nothing leaves the land empty. Everything stays in motion - circular, regenerative, and self-renewing.

Leading without mining also means leading without dependency. Mining ties economies to volatile markets and political influence. Hydrogen, in contrast, ties nations to knowledge, technology, and cooperation. It creates industries that grow through innovation, not extraction. It rewards education, not exploitation.

This shift is more than environmental; it is moral. Africa's leadership in clean energy must reflect a new kind of wisdom - the wisdom to create prosperity without harm. The continent can show the world that growth and peace can coexist, that technology can heal the planet instead of exhausting it.

GERD+ represents this philosophy in action. By using water and light to generate hydrogen and ammonia, it proves that Africa can power its future through creation rather than destruction. The reservoirs, solar fields, and hydrogen hubs become living infrastructures - systems that restore rather than remove.

This is also a leadership of example. While much of the world still depends on mining to build renewable technologies - for lithium, cobalt, or nickel - Africa can demonstrate an alternative path: energy from natural cycles, not from scars in the ground. By combining hydropower, solar, and circular systems, Africa can prove that clean energy does not require dirty methods.

Leading without mining is not about rejecting industry; it is about redefining it. Africa can still manufacture, export, and innovate - but with resources that renew themselves. The true value of leadership lies not in what we take, but in what we give back.

When Africa stands before the world as a hydrogen-powered continent, it will do so with clean hands and clear conscience. It will not carry the weight of extraction but the lightness of creation.

The era of digging must give way to the era of designing. Africa's strength will come not from what it pulls from the ground, but from what it brings to life. The continent that once powered empires through its minerals will now power humanity through its wisdom.

3.4 The Case for Green Hydrogen: Water, Sun, and Intelligence

Every great civilization is built upon three forces: natural resources, knowledge, and will. For Africa, these three come together in the perfect trinity of **water, sun, and intelligence**. This is the foundation of green hydrogen - the clean energy that can power not only our industries but also our independence.

Water is life. It is movement, rhythm, and potential. Within every drop lies the energy that can light a home or drive a machine. Through electrolysis, water becomes the source of hydrogen - the simplest, cleanest fuel known to humankind. Africa's rivers, lakes, and reservoirs hold vast potential, enough to sustain a continental hydrogen system. When used wisely, this water can serve both humanity and nature, creating energy without diminishing life.

The sun is abundance. It shines freely over the continent, demanding no payment and showing no borders. Africa receives more solar radiation than any other landmass on Earth. Yet, much of it still goes unused. By converting sunlight into electricity, and electricity into hydrogen, the continent can transform an invisible gift into visible progress. In the GERD+ model, sunlight is not wasted on the surface of reservoirs; it is captured by floating solar systems that power the electrolyzers below. Water and light work together as they always have in creation - one reflecting, one generating, both sustaining.

Intelligence is the third and most essential ingredient. It is what turns potential into power. Without intelligence, resources remain silent. With it, they speak and multiply. Intelligence means not only scientific knowledge but also strategic wisdom - the ability to plan, integrate, and protect. Africa's greatest resource is not its minerals or its rivers; it is the brilliance and creativity of its people. When guided by education, innovation, and cooperation, this intelligence can convert the continent's natural abundance into enduring strength.

The case for green hydrogen is therefore both practical and moral. Practically, it is the only pathway to sustainable industrialization. It allows Africa to produce fertilizers for its farmers, fuels for its transport systems, and power for its industries - all without carbon, without pollution, without dependence. Morally, it represents a new relationship with the planet: one where we generate prosperity without extraction, and progress without destruction.

Unlike fossil fuels, green hydrogen never runs out. It does not poison the air or divide nations through scarcity. It brings stability, cooperation, and peace. It allows countries to trade not just goods, but goodwill - sharing energy through clean pipelines and shared technologies rather than competing over limited reserves.

The **GERD+** vision embodies this logic. It links hydropower with sunlight and transforms both into hydrogen, which in turn becomes ammonia, fertilizer, and fuel. It turns one dam into an ecosystem - a living network of energy and renewal. It shows that intelligence is not about having the most resources, but about connecting them in the most natural and human way.

Africa's case for green hydrogen is not an argument; it is evidence. The continent already holds everything it needs: abundant water, endless sunlight, and the minds to unite them. What remains is coordination - the will to build systems that reflect the rhythm of the land and the aspirations of its people.

When water, sun, and intelligence move together, they form a single current. That current is Africa's future. And its colour - pure, renewable, and powerful - is green.

3.5 How Hydrogen Can Power Africa's Industrial Future

Every generation inherits two choices: to follow the world or to redesign it. Africa's generation today stands before that choice. Hydrogen gives the continent a chance to redesign - to build industries that are clean, modern, and entirely African in their foundation.

Hydrogen is not only an energy source; it is a raw material for transformation. It can fuel ships and factories, produce fertilizers, refine metals, and power heavy transport. It can replace the coal, oil, and diesel that once defined industrial power, allowing Africa to leap directly into a green economy. In doing so, it can build the industries the continent has long needed - industries that create jobs, export value, and preserve the environment all at once.

In the **GERD+** model, hydrogen is the bridge between energy and economy. Electricity from hydropower and floating solar systems is converted into hydrogen and ammonia. These molecules then become the building blocks of new industries:

- **Green fertilizers** for agriculture, reducing dependency on imported chemicals.
- **Clean fuels** for trucks, trains, and ships that move goods across the continent.

- **Hydrogen-powered manufacturing** for steel, cement, and industrial gases.
- **Export corridors** that send green hydrogen and ammonia to global markets through African ports.

This is how Africa can industrialize without repeating the mistakes of others - without polluting its air, draining its soils, or dividing its people. Hydrogen turns Africa's rivers and sunlight into the new oil of the 21st century - but an oil that heals rather than harms.

Industrialization through hydrogen also means job creation on every level. Engineers, technicians, scientists, builders, and farmers all become part of the same ecosystem. Hydrogen plants require skilled labour for design, construction, and operation. Supply chains create opportunities for logistics, training, and innovation. Education systems grow alongside industries, preparing a generation not just to work, but to lead.

The hydrogen economy is also an economy of inclusion. It connects rural and urban areas, highlands and coasts, energy producers and consumers. For example, ammonia made at GERD+ can be used as fertilizer in farms across East Africa, powering food production. The same molecule can then fuel the ships that transport that food through the Pan-African Canal to other regions. One product, one system, many benefits - all within Africa.

Hydrogen also strengthens continental cooperation. No single country can manage the entire hydrogen chain alone. It requires pipelines, ports, grids, and shared standards. This necessity for partnership naturally pushes nations to work together, turning energy into diplomacy. The Pan-African Hydrogen Corridor - envisioned alongside the Pan-African Canal - is not just an infrastructure project; it is a pathway to unity through shared prosperity.

Most importantly, hydrogen gives Africa ownership of its destiny. Instead of exporting raw materials and importing finished products, the continent can now export finished energy and finished goods. It can set its own value, negotiate its own terms, and build an economy that serves its people first.

Hydrogen is the missing link between Africa's natural wealth and its industrial future. It carries the potential to turn water into wages, sunlight into salaries, and knowledge into national pride.

The real power of hydrogen lies not in its flame, but in what it ignites within us - the confidence that Africa can industrialize in its own way, on its own terms, and for its own people.

When that confidence meets cooperation, Africa's industries will no longer chase the world; they will light the way for it.

4.0 BUILDING THE GERD+ SYSTEM

A great idea becomes real only when it is built. GERD+ is more than a concept; it is a continental design that can be realized step by step, layer by layer, with precision and purpose. It combines engineering, ecology, and economics into a single living framework.

Building GERD+ does not mean constructing a new dam. It means **expanding the role of the existing Grand Ethiopian Renaissance Dam (GERD) into a multi-dimensional energy ecosystem**. Each part - hydropower, floating solar, hydrogen, ammonia, and storage - connects naturally to the next, forming a continuous cycle of production, transformation, and renewal.

At the heart of the system lies the dam itself, generating clean hydropower around the clock. Above its reservoir float solar panels that capture sunlight during the day, reducing evaporation and feeding electricity into the electrolysis plants on the shore. These plants split water into hydrogen and oxygen. The hydrogen is then used directly as clean fuel or synthesized into ammonia with nitrogen from the air.

To balance supply and demand, the system includes underground hydrogen storage in the **Afar salt caverns** - a natural geological formation ideal for secure, large-scale storage. The stored hydrogen can be released when electricity demand peaks, ensuring stability and reliability across the regional grid.

From the Afar hub, pipelines extend through the **Pan-African Energy Corridor**, connecting Ethiopia to neighbouring countries and to sea ports for export. Along these routes, industrial and agricultural hubs emerge - fertilizer plants, clean logistics centers, hydrogen refuelling stations, and training institutes. Each link of the chain adds value and employment, turning infrastructure into livelihood.

But the most important aspect of building GERD+ is **integration**. Every element supports the others:

- Hydropower provides base energy.
- Floating solar adds extra capacity and conserves water.
- Hydrogen stores excess power and stabilizes the grid.

- Ammonia expands trade and agriculture.
- Salt caverns secure long-term reserves.

This interconnected design transforms GERD+ into a circular system, where no energy is wasted and every process reinforces another. It embodies the African principle of wholeness - that life thrives when all parts support the whole.

The construction of such a system requires vision and coordination more than heavy machinery. It needs cooperation among ministries, engineers, financiers, and educators. It calls for policies that encourage shared ownership and local capacity-building. It also demands trust - trust that African hands and minds can design and operate one of the most advanced energy systems in the world.

GERD+ is not a single project but a **template** - a blueprint that can be replicated across Africa's other rivers and reservoirs. From the Congo Basin to the Zambezi, similar integrated systems can form a network of clean power, making the continent self-sufficient in energy and food.

Building GERD+ is about more than producing electricity. It is about building confidence, capability, and cooperation. It shows that when Africa constructs its systems around nature's logic - not against it - every structure becomes sustainable.

When complete, GERD+ will stand not only as an engineering marvel but as a symbol of how Africa builds when it builds with wisdom. It will show the world that the future of energy is not extraction, but integration. And that the future of Africa is not waiting - it is being built, piece by piece, with light, water, and hope.

4.1 The Energy Triangle: Hydropower, Solar, and Hydrogen

Every strong system is built on balance. In nature, day follows night, rain follows drought, and rivers meet the sea. In the same way, a resilient energy system must balance its sources - combining stability, abundance, and adaptability. This harmony is at the heart of **GERD+**, which rests upon what can be called **the Energy Triangle**: **hydropower, solar power, and hydrogen**.

Each corner of this triangle plays a distinct role, and together they create a living energy cycle that mirrors the rhythm of the continent itself.

Hydropower: The Steady Heartbeat

Hydropower is the anchor of the system - strong, constant, and dependable. It provides the foundation on which all else stands. The Grand Ethiopian Renaissance Dam represents this strength. It operates day and night, through seasons and storms, delivering reliable electricity to homes and industries. Its reservoirs store not only water but also potential - the stored energy of Africa's renewal.

Hydropower embodies continuity. It is what keeps the system alive when sunlight fades. It supplies stability to the grid and allows flexibility in managing energy flows. It also provides the vital electricity that powers the electrolysis process - the bridge between water and hydrogen.

Solar Power: The Infinite Partner

Where hydropower is constant, solar power is abundant. The sun provides more energy to Africa each hour than the world consumes in a year. Floating solar panels on the GERD+ reservoir extend this potential - absorbing sunlight, reducing evaporation, and doubling the system's capacity without adding new dams or displacing land.

Solar power fills the daylight hours, charging electrolyzers to produce hydrogen and powering local industries. At sunset, hydropower takes over, ensuring an uninterrupted supply. The two systems complement each other perfectly, like breath in and breath out - one shining, one flowing, both alive.

Hydrogen: The Bridge and the Brain

Hydrogen completes the triangle. It connects generation to storage, and storage to industry. It is the translator between renewable energy and human needs. When water is split by electricity, hydrogen carries the power forward - into factories, transport, and agriculture. It can be stored for months, exported across oceans, or turned into ammonia for fertilizer and fuel.

Hydrogen is the bridge that transforms the physical into the practical - turning light and flow into livelihoods. It gives flexibility to the system, allowing energy to move across time and space. When the sun and rivers produce more than needed, hydrogen stores it. When demand rises, it gives it back. In this way, the triangle becomes a living loop - continuous, adaptive, and circular.

The Harmony of the Three

Hydropower brings **stability**, solar brings **abundance**, and hydrogen brings **mobility**. Together, they make energy flow like an ecosystem - dynamic but balanced, powerful but peaceful.

This triangle can be replicated anywhere in Africa. Every country with sunlight, water, and will can form its own version of the GERD+ system - scaled to its geography and needs. When these systems connect across borders through the **Pan-African Energy Network**, the continent itself becomes one great triangle, harmonizing nature and technology.

The triangle also symbolizes unity - three points supporting one another, none dominant, all essential. It represents the African way of balance: not the competition of forces, but their cooperation.

When Africa learns to manage energy as nature manages life - through balance, renewal, and flow - no darkness will last. The triangle of hydropower, solar, and hydrogen is more than technology; it is a philosophy of harmony in motion.

4.2 The Afar Salt Caverns: Africa's Energy Storage Hub

Every powerful system needs a place to rest - a heart that beats and a lung that breathes. For **GERD+**, that breathing space is found beneath the earth, within the natural architecture of Ethiopia's **Afar region**, where ancient salt formations create one of the most remarkable geological treasures on the continent. These are the **Afar salt caverns**, and they are designed by nature to become Africa's first great energy storage hub.

Hydrogen, though light and abundant, must be stored safely and efficiently to unlock its full potential. The world's most advanced energy systems - from Europe to North America - are already turning to **salt caverns** as their preferred solution. These caverns, formed through the slow movement of salt over millions of years, are airtight, self-healing, and stable under high pressure. They can hold massive quantities of hydrogen gas without leakage or contamination, making them ideal for long-term, large-scale storage.

The **Afar salt basin** in Ethiopia's Danakil Depression offers one of the best geological locations for such a system. Stretching across a vast area shaped by volcanic and tectonic forces, its salt beds are thick, deep, and pure. The same forces that once formed salt lakes and geysers now provide a natural vault for the continent's future energy.

In the **GERD+** system, the process is simple yet powerful. Excess electricity from the dam and floating solar panels drives electrolysis, producing hydrogen. When production exceeds immediate demand, that hydrogen is compressed and injected into the Afar salt caverns. There, it rests safely - stored energy waiting to be used. When the grid requires extra power or when ships and industries need supply, the stored hydrogen is released, converted back into electricity, or transported through pipelines toward the Pan-African Canal corridor.

This approach turns hydrogen from a fleeting resource into a permanent asset. It transforms Ethiopia's geology into an integral part of the continental energy cycle. The Afar caverns become the lungs of Africa's renewable system - inhaling energy during times of surplus, exhaling it when needed.

The benefits are enormous:

- **Energy stability:** Africa gains a natural battery capable of storing weeks or even months of renewable power.
- **Economic resilience:** Stored hydrogen ensures stable supply and prices for industries and export partners.
- **Strategic security:** Energy reserves underground provide protection against droughts, demand shocks, or system interruptions.
- **Continental cooperation:** Neighbouring nations can share storage and draw from it during need, strengthening unity and reliability.

Beyond technology, the Afar hub is symbolic. It represents a deeper lesson about Africa's future - that we do not need to destroy nature to build progress. Instead, we can work with nature's designs, respecting its patterns and turning its quiet strength into lasting value.

The Afar salt caverns also have the power to inspire a new kind of pride. They show that Africa's ancient lands, often seen only for their beauty or tourism, hold the keys to the most advanced energy systems of the future. What was once seen as barren desert now becomes the core of continental innovation.

In a single system, GERD+ connects the surface and the depths - light from the sun, flow from the river, and silence from the underground. Together they form a complete cycle: **generation, transformation, and storage**.

When the first hydrogen caverns of Africa are sealed in the Afar Basin, it will mark not just an engineering milestone, but a spiritual one - the moment Africa learns to breathe through its own earth.

4.3 The Pan-African Pipeline: Connecting Energy, Ports, and People

Energy is powerful only when it moves. The strength of GERD+ does not lie in one dam, one reservoir, or one nation - it lies in connection. To unite Africa's energy and trade systems, the continent needs arteries that carry not just power, but purpose. This is the vision behind the **Pan-African Pipeline**, a living corridor designed to connect **energy, ports, and people**.

The pipeline is more than a structure of steel and valves. It is the nervous system of the **GERD+** model - the physical link that allows hydrogen, ammonia, and clean electricity to travel from Ethiopia's highlands to Africa's coasts, from the interior to the world. Like the **Pan-African Canal**, it forms part of a greater continental network that binds regions together in shared progress.

The design follows a simple and elegant logic. Hydrogen produced at GERD+ and stored in the Afar salt caverns travels through a high-efficiency pipeline that runs along a multi-purpose corridor - parallel to future canal routes and railways - reaching ports on the Red Sea and the Gulf of Aden. From there, the clean molecules can be exported as **green hydrogen or ammonia**, powering ships, industries, and nations far beyond Africa's shores.

But export is only one function. The pipeline also carries energy to **domestic markets**, fuelling African industries, fertilizer plants, and transport systems along the way. Every section of the route becomes a catalyst for growth - a hub of employment, education, and innovation. Towns grow where stations are built, farms thrive where fertilizer becomes affordable, and knowledge circulates where energy flows.

The Pan-African Pipeline is, at its core, a peace project. It creates **mutual dependency through shared benefit**. No country along the route stands alone; each one contributes to and gains from the system. When energy and prosperity move across borders, conflict loses its purpose. Cooperation becomes not a choice but a necessity.

The pipeline is also designed for **adaptability**. It can carry both hydrogen and ammonia, or later be expanded to include synthetic fuels or captured carbon streams. It can connect to solar and hydropower hubs across Africa, forming an integrated **Pan-African Energy Belt** that links east to west and north to south. Over time, smaller pipelines will branch outward, serving local cities, villages, and industries - until the continent itself becomes a continuous web of clean energy flow.

From an engineering perspective, the system builds on proven technologies used worldwide. Modern hydrogen pipelines already operate safely in many countries. The Afar storage hub and GERD+ production complex provide the steady supply needed to justify such an investment. But the African innovation lies not in copying, but in combining - aligning infrastructure with geography, trade, and social development.

Each kilometer of the Pan-African Pipeline carries more than molecules. It carries dignity. It turns Africa's geography - once seen as a barrier - into a bridge. It connects landlocked nations to the sea, rural communities to modern markets, and youth to employment in the industries of tomorrow.

In spirit, the pipeline is not made of metal; it is made of vision. It embodies the same principle that defines all living systems: interdependence. Just as rivers feed farms and trees feed air, energy must feed the people who produce it.

The Pan-African Pipeline is the pathway through which Africa's light will travel - quietly, efficiently, and forever forward. It will not only connect ports and people, but also ideas and futures. It will prove that when Africa builds for unity, its power multiplies.

4.4 Integration with the Pan-African Canal System

Water has always been the lifeline of civilization. The great cities of history - from the valleys of the Nile to the banks of the Niger - were built where water moved. In the twenty-first century, Africa has the chance to reclaim that wisdom by making water not only the foundation of agriculture and transport, but also of clean energy. This is the deeper meaning behind the **integration of GERD+ with the Pan-African Canal System**.

The **Pan-African Canal**, a visionary concept created by the author, is not merely a waterway. It is a continental artery - designed to give sea access to landlocked nations, irrigate farmlands, and create a new era of green logistics and food security. When integrated with **GERD+**, it becomes something even more profound: a **hydrogen-enabled water corridor**, where navigation, irrigation, and energy exist as one circular system.

The logic is natural and elegant. The same canal that carries ships and water can also host energy pipelines and floating solar fields. The same flow that feeds crops can feed electrolysis plants. Every element strengthens the others - water for transport, sunlight for energy, hydrogen for fuel, and ammonia for fertilizer. This is the physical embodiment of a *living system*, where nothing stands alone.

In this model, **GERD+** serves as a primary energy source for the canal's infrastructure. Its hydropower and hydrogen production capacity can supply electricity to canal locks, irrigation systems, and logistics hubs across multiple countries. In return, the canal offers the GERD+ system a transportation route - a pathway for exporting clean hydrogen and ammonia to global markets and distributing fertilizers within Africa.

Imagine cargo ships moving silently through the canal, powered not by diesel but by green hydrogen produced at GERD+. Imagine ports illuminated by clean electricity and surrounded by industries that use the same energy to create African-made goods. Imagine farms along the canal banks using locally produced ammonia fertilizers to grow food that sustains millions. This is not a dream; it is a design waiting to be realized.

The **integration between GERD+ and the Pan-African Canal** also carries a message of unity. It links the heart of the continent to its coasts - from the highlands of Ethiopia to the harbours of the Red Sea and beyond. It transforms national projects into continental systems. It replaces isolation with flow, fragmentation with coherence. Every participating nation becomes a stakeholder in the same network of prosperity.

From an ecological perspective, this integration brings balance. By using floating solar fields along canal surfaces, the system can generate electricity while reducing water evaporation. By transporting hydrogen through pipelines running parallel to the canal, the system minimizes emissions and prevents reliance on fossil-fuelled transport. By combining irrigation, trade, and energy, it ensures that growth in one sector nourishes the others.

The Pan-African Canal becomes, therefore, more than infrastructure - it becomes **a symbol of Africa's unity through circulation**. It demonstrates that development can move like nature itself: interconnected, regenerative, and abundant.

GERD+ and the Pan-African Canal together show the power of design when it mirrors life. A dam that produces energy, a canal that carries water and trade, and a network that binds them both through hydrogen and sunlight - this is the architecture of a living continent.

Integration does not simply connect systems; it connects stories. The story of energy meets the story of food, the story of water meets the story of movement, and together they form the story of Africa's transformation.

The canal and GERD+ are two visions born from the same belief: that Africa's greatness lies not in imitation, but in imagination. United, they turn that imagination into motion - a continental river of light, water, and purpose flowing toward a shared future.

4.5 From Power Generation to Energy Security

Generating power is only the beginning. True transformation comes when that power becomes **security** - when it protects livelihoods, stabilizes economies, and strengthens sovereignty. **GERD+** was never imagined as merely a generator of electricity. It was designed as a foundation for Africa's long-term energy security, independence, and peace.

Energy security means that no African nation should face darkness when it has sunlight. It means no farmer should suffer crop loss when water and power are near. It means no country should depend on imported fuel when it holds enough natural potential to sustain itself and its neighbours. GERD+ transforms this principle into practice.

At its core, the system ensures **continuity**. Hydropower provides reliable base load. Floating solar panels extend generation capacity. Hydrogen and ammonia storage guarantee that energy can be delivered even when weather or demand fluctuates. This integration shields the continent from shortages, price shocks, and the instability of global fuel markets.

Energy security also means **diversification**. Africa's current dependence on imported diesel and coal exposes it to external risk. By expanding clean hydrogen, ammonia, and solar-based grids, GERD+ diversifies not only sources but also uses - supplying electricity, fuel, and fertilizer through one connected network. Each element strengthens the others: if one system slows, another sustains. This is resilience by design.

But the deepest meaning of security is **freedom**. For too long, Africa's economies have been shaped by foreign energy systems - contracts, refineries, and grids designed elsewhere. GERD+ reverses that dependency. It keeps ownership of both resources and technology within Africa. The revenue from energy sales funds education, innovation, and infrastructure, not debt. The skills developed to operate the system remain in African hands.

Energy security must also include **food security**. The ammonia produced from hydrogen becomes fertilizer for African farms, ensuring self-sufficiency in agriculture. The same molecule that fuels ships and industries feeds crops and people. Through this circular relationship, energy becomes nourishment, and nourishment becomes stability.

In addition, the GERD+ system promotes **regional cooperation**, which is the highest form of security. Shared energy creates shared interest. When nations trade power, they trade peace. The flow of electricity, hydrogen, and ammonia across borders transforms neighbours into partners. The interconnected grids and pipelines of GERD+ make conflict less possible and unity more practical.

Environmental stability is another pillar of energy security. GERD+ achieves it naturally. It generates no carbon, leaves no pollution, and restores balance to the atmosphere. By using water and sunlight instead of fossil fuels, Africa protects its ecosystems while powering its growth. Security, therefore, is not only about protection from threats but also preservation of harmony with nature.

Through GERD+, Africa moves from **generation to guardianship** - from producing energy to protecting its people and its planet. The system ensures that power serves life, not the other way around. It transforms vulnerability into vitality.

The result is not just energy independence, but **civilizational security** - the confidence that Africa can stand, sustain, and shine on its own terms. Energy becomes more than electricity; it becomes assurance. It becomes the quiet strength that allows nations to dream without fear.

When a continent controls its light, it controls its future. GERD+ is that control - the promise that Africa's power will never again depend on permission, but on purpose.

5.0 AFRICA'S ENERGY PORTFOLIO

Every continent has a portfolio - a collection of its most valuable assets. For Africa, that portfolio is not held in banks or stock exchanges; it is held in rivers, sunlight, land, and people. Energy is the foundation of that wealth, and how Africa manages it will define its future.

An **energy portfolio** is more than a list of power plants. It is a strategic balance between generation, storage, and use - between what nature gives and what society needs. It is how nations choose which projects to build, which technologies to prioritize, and which systems to connect. The strength of a portfolio lies not in the number of projects it contains, but in how well they complement each other.

Africa's challenge is not a lack of resources but a lack of coordination. Dozens of nations, hundreds of projects, and billions of dollars often move without a shared framework. The result is fragmentation - islands of light in a sea of darkness. The purpose of GERD+ and its sister initiatives, such as the **Pan-African Canal**, is to turn those scattered lights into a single, shining network.

An African energy portfolio must be built around five guiding principles:

- **Diversity** – combining hydropower, solar, geothermal, and hydrogen.
- **Integration** – linking systems across borders.
- **Circularity** – ensuring that waste from one process fuels another.
- **Self-reliance** – reducing dependence on imported fuels and technologies.
- **Sustainability** – balancing human needs with ecological harmony.

GERD+ fits at the core of this continental portfolio. It is not one project among many; it is the model that demonstrates how all can work together. It shows that energy systems should not compete - they should complement. Where the sun is strong, build solar. Where rivers flow, harness hydro. Where coastlines meet trade routes, develop hydrogen and ammonia export terminals. The portfolio thrives when every nation contributes according to its strengths.

The goal is to manage energy as a **continental asset**. This means creating cross-border energy corridors, shared grids, and coordinated investment plans. Africa's rivers, deserts, and coasts are not national boundaries but collective opportunities. Just as Europe integrated its coal and steel to build peace, Africa can integrate its light and water to build prosperity.

Energy portfolio management also requires **visionary leadership** - leaders who can think beyond electoral cycles and see generations ahead. GERD+ and related systems must be governed not by short-term contracts but by long-term purpose: to create value that endures, jobs that uplift, and technologies that belong to Africa.

The portfolio approach also brings resilience. When one region faces drought, another's solar power can compensate. When one grid experiences shortage, stored hydrogen elsewhere can stabilize supply. Energy becomes not a national issue but a continental insurance system - a safety net woven by cooperation.

Finally, Africa's energy portfolio is not just physical; it is philosophical. It reflects how the continent sees itself - not as a recipient of technology but as a creator of civilization. When Africa builds its energy systems with its own logic and values, it defines not only its economy but its identity.

The time has come to manage Africa's power as Africa's own. GERD+, the Pan-African Canal, and other transformative projects are not isolated dreams - they are the first entries in a living portfolio of light.

The day Africa learns to manage its energy as a collective investment, the continent will no longer speak of shortages or dependence. It will speak of abundance, security, and unity - a portfolio not of numbers, but of hope turned into power.

5.1 Energy as a Continental Portfolio

Energy is the bloodstream of a living continent. When it flows freely, life thrives - industries grow, food is produced, and knowledge moves. When it is blocked, nations weaken, and opportunity fades. For Africa to live as one, it must treat its energy not as scattered projects, but as a **continental portfolio** - shared, coordinated, and strategically balanced across its vast geography.

In a single company, a portfolio represents diversity and resilience. No wise investor places all value in one product or one market. In the same way, no nation should depend on a single energy source, and no continent should depend on one region's abundance while others struggle in scarcity. A continental portfolio means every African country contributes its strength and benefits from the whole.

Hydropower in Ethiopia, solar energy in the Sahel, geothermal in the Rift Valley, wind along the coasts, and biomass in equatorial zones - each plays a unique role in the continental mix. But coordination is the difference between potential and prosperity. If managed as one system, Africa's natural resources can sustain all its people and export surplus energy to the world.

The **GERD+** concept offers a blueprint for this coordination. It demonstrates how energy can be produced, stored, and traded through interconnected systems - hydropower feeding solar grids, hydrogen pipelines linking inland regions to ports, and ammonia exports financing local growth. The model converts isolated assets into an integrated network, where the success of one project strengthens all others.

A continental portfolio also means shared standards, shared infrastructure, and shared vision. Instead of each nation building its own energy islands, Africa can develop regional clusters - East African hydropower and hydrogen hubs, West African solar and gas transition zones, North African export corridors, and Southern African industrial basins. Together, these become the building blocks of a continental energy union.

This approach brings both efficiency and equity. Wealthier nations can invest in research and manufacturing, while resource-rich nations supply the raw potential. Landlocked countries gain access through regional pipelines and canals, ensuring that geography no longer defines destiny. When energy flows across borders, prosperity follows.

The African Union's Agenda 2063 envisions an interconnected, prosperous continent. A continental energy portfolio turns that vision into structure. It gives Africa the capacity to plan as one body - forecasting demand, managing reserves, and distributing benefits fairly. In this system, energy is not a commodity but a shared right.

A portfolio mindset also encourages flexibility. If drought affects hydropower, solar fields fill the gap. If sunlight dims, hydrogen storage releases reserves. The system adapts like nature itself - shifting, balancing, and renewing. This resilience is the hallmark of a living portfolio.

But the greatest value of a continental portfolio is **unity**. It builds interdependence that discourages division. When electricity, hydrogen, and trade flow between nations, borders lose their power to separate. Cooperation becomes not only moral but practical.

Managing energy as a continental portfolio means treating Africa's light, water, and knowledge as common wealth - not owned by one, but shared by all. It transforms energy from a national ambition into a continental identity.

The future will not belong to the richest country, but to the most connected continent. Africa's wealth is not in what it owns individually, but in what it builds collectively. GERD+ is the first piece of that continental portfolio - the heart that pumps life through Africa's arteries of light.

5.2 Regional Energy Hubs: From the Nile to the Niger

Africa is a continent of rivers, deserts, and coasts - each with its own rhythm, each carrying a different kind of power. To manage this vast and varied landscape, energy must be organized not through competition but through **coordination**. The future of African energy lies in creating **regional energy hubs**, connected through shared corridors of power, trade, and knowledge. From the Nile to the Niger, from the Congo to the Zambezi, each region can become a pillar of a living continental grid.

These hubs are not simply technical centers. They are ecosystems of progress - places where generation, industry, research, and community development grow together. They turn geography into opportunity, making every part of the continent valuable in its own way.

The Nile Hub – Water, Light, and Innovation

The **GERD+ system** anchors the Nile hub. Here, Ethiopia's hydropower and floating solar capacity merge to create the world's most advanced renewable-energy ecosystem. It powers hydrogen and ammonia production for agriculture and industry, while the **Afar salt caverns** provide strategic storage. From this base, energy flows north and west through shared grids and pipelines, connecting to Sudan, South Sudan, and Egypt, and eastward through the Pan-African Canal toward the Red Sea.

The Nile hub symbolizes Africa's awakening - a reminder that the river which once sustained ancient civilizations can now power a modern one.

The Niger Hub – Solar and Hydrogen in the Sahel

Stretching across Mali, Niger, and Nigeria, this region receives some of the highest solar radiation on Earth. The **Niger Hub** can become the beating solar heart of Africa - an expanse of sunlight turned into hydrogen through desert electrolysis plants. This clean hydrogen can be transported through pipelines to North African ports or eastward toward the Pan-African Energy Belt. Local industries, powered by renewable energy, can manufacture fertilizers and chemicals for regional markets, ensuring food and energy security.

In this hub, the desert itself becomes productive - a field of light turning emptiness into abundance.

The Congo Hub – The Forest Energy Balance

In Central Africa, the **Congo Basin** offers unmatched hydropower potential through the Inga Falls and other river systems. The region can host massive renewable plants while preserving its rainforests, the lungs of the world. Through proper management, Congo's water can generate electricity and hydrogen for the entire region while maintaining its ecological balance.

The Congo Hub will also connect to forestry and **circular economy programs**, aligning renewable energy with carbon absorption and ecosystem protection - an example of how progress can live in peace with nature.

The Southern Hub – Industrial Transition Zone

In Southern Africa, countries like Zambia, Zimbabwe, Botswana, and South Africa already have developed industrial infrastructure. The Southern Hub will focus on transforming these existing industries to run on hydrogen instead of coal or diesel. Hydrogen-based steel production, green transport systems, and solar-powered manufacturing can redefine Africa's industrial landscape while cutting emissions.

This hub becomes the engine room of Africa's green transformation - turning the factories of the past into the factories of the future.

The Eastern and Coastal Hub – The Gateway of Trade

Along the eastern corridor, from Kenya and Tanzania to Mozambique, the **Pan-African Canal** and coastal hydrogen terminals will link inland production centers to global markets. These ports will export hydrogen and ammonia produced at GERD+ and other hubs, while importing technology, knowledge, and investment. They will also serve as centers for hydrogen-powered maritime logistics, building Africa's capacity for green shipping.

The eastern hub becomes Africa's open gate - a bridge between inland energy and international exchange.

Each hub represents a chapter in the same story - different voices in one harmony. Together, they create a network that can balance supply and demand across the continent, stabilize grids, and promote shared prosperity.

When energy flows from the Nile to the Niger, Africa will no longer be a map of separate nations but a living system of interlinked strengths. The river, the desert, the forest, and the sea - all become parts of one continuous pulse.

This is how Africa grows: not by copying models from elsewhere, but by linking the natural rhythm of its regions into one heartbeat of light.

5.3 Distributed Power and National Grids

Africa's vastness is both its challenge and its gift. Villages, towns, and cities stretch across deserts, mountains, and forests, often separated by hundreds of kilometers. In such a landscape, a single centralized grid cannot serve everyone. To light every home and power every community, Africa must combine **national grids** with **distributed power systems** - a hybrid model where big and small, central and local, coexist in balance.

A traditional grid is like a great river - powerful but distant. A distributed network, by contrast, is like rainfall - small streams of energy falling everywhere, nourishing the land directly. When these two systems are connected, a continent can have both reach and resilience.

The **GERD+** framework is designed for exactly this kind of balance. The main hydropower and solar-hydrogen systems generate large-scale electricity for transmission across countries, but part of that power is also converted into hydrogen and ammonia to fuel smaller, decentralized systems - **mini-grids, micro-factories**, and **local industries**. These smaller systems can operate independently when needed, ensuring that even the most remote areas remain powered.

Distributed Power: Energy for Everyone

Distributed power means giving communities ownership over their light. Small solar farms, hydrogen micro-grids, and bioenergy systems can operate in villages far from national infrastructure. They can power schools, clinics, and small businesses, enabling development where large transmission lines have yet to reach.

By decentralizing production, Africa not only expands access but also builds **resilience**. When one part of a national grid experiences disruption, the distributed systems continue to function, ensuring that hospitals, farms, and industries never lose power. This creates stability in times of drought, conflict, or natural disaster.

Distributed systems also generate local employment. Technicians, electricians, and operators are needed in every region, spreading economic opportunity along with energy. The same sunlight that powers a dam can power a village; the same hydrogen pipeline that fuels ships can supply local industries.

National Grids: The Backbone of Integration

While distributed systems bring access, **national grids** bring connection. They remain the backbone of industrial growth and large-scale coordination. Every hub, from the Nile to the Niger, must feed into a unified grid that allows nations to trade electricity and balance supply.

Through **interconnection**, countries can share their strengths: Ethiopia's hydropower stabilizing Kenya's solar output, or Nigeria's solar surplus supporting its neighbours. The more connected the grids, the less vulnerable each nation becomes to local shortages or weather fluctuations.

GERD+ adds another layer to this design through **hydrogen-based storage and balancing**. Excess energy from national grids can be converted into hydrogen during off-peak hours and released when demand rises. This chemical storage creates a flexible, long-lasting alternative to battery systems, ensuring that power is always available when needed.

The African Energy Web

The goal is to create an **African Energy Web** - a network of national grids linked by regional corridors, strengthened by thousands of local distributed systems. In this web, energy flows like information: freely, intelligently, and continuously.

Each country becomes both a producer and a consumer, each region both a hub and a link. This cooperative framework transforms competition into collaboration and scarcity into stability.

The African Energy Web will not only light cities but also connect hearts - a practical symbol of the continental unity envisioned by Agenda 2063. It will demonstrate that when power is shared wisely, it multiplies; when it is hoarded, it fades.

In this way, Africa's grid of the future will not resemble those of the past - fragile, centralized, and exclusionary. It will be a **living grid**, made of millions of lights working together, from the turbines of GERD+ to the solar rooftops of distant villages.

Energy, like knowledge, becomes strongest when it is distributed. And in Africa, it will flow not from center to edge, but from everywhere to everyone.

5.4 Clean Logistics and Hydrogen Corridors

Transport is the bloodstream of civilization. Without movement, there is no trade, no connection, no growth. Yet for too long, Africa's movement has been powered by fuels that pollute the air, drain foreign reserves, and weaken economic independence. The next chapter of Africa's progress must therefore be built on **clean logistics** - transport systems powered not by diesel, but by **hydrogen**.

The **hydrogen corridor** is the backbone of this transformation. It connects the continent's inland energy producers with its coastal trade routes, industrial zones, and urban centers. It links the **GERD+ system** and other hydropower-solar hubs to the **Pan-African Canal** and the **Pan-African Pipeline**, forming one continuous flow of clean energy and mobility.

Hydrogen as the New Fuel of Movement

Hydrogen is perfectly suited for Africa's vast distances. Unlike batteries, which are best for short-range use, hydrogen can power long-haul trucks, trains, and ships without heavy storage or long charging times. It can also be converted into **green ammonia**, which is safer to transport and can serve as a global shipping fuel.

In the GERD+ model, hydrogen is produced from renewable energy and delivered through pipelines that run alongside transport corridors - the same paths used for highways, railways, and canals. This integration ensures that logistics and energy infrastructure grow together, reducing costs while maximizing reach.

Imagine freight trains running across the continent on hydrogen fuel cells. Picture buses and trucks refuelled at hydrogen stations built along the Pan-African transport network. Visualize ships sailing from African ports, powered by clean ammonia produced inland by African engineers. Every movement becomes clean, quiet, and independent.

The Pan-African Hydrogen Corridors

The vision extends across several strategic corridors:

- **The East African Corridor**, stretching from the GERD+ complex through Kenya, Tanzania, and Mozambique, connecting inland hydrogen production to coastal ports and shipping routes.
- **The Sahelian Solar Corridor**, running from Niger and Mali to North Africa, where solar hydrogen plants feed export terminals along the Mediterranean.
- **The Central Energy Corridor**, linking the Congo Basin to Southern Africa's industrial hubs.
- **The Pan-African Canal Corridor**, where floating solar, hydrogen barges, and ammonia transport vessels integrate waterborne trade with renewable energy systems.

Each corridor functions as a **clean logistics spine**, carrying energy, goods, and knowledge. Together, they transform Africa into a self-sustaining trade network - efficient, affordable, and carbon-free.

Local Benefits and Industrial Growth

Clean logistics are not only for export. They improve everyday life. Hydrogen-powered buses can provide affordable public transport in cities. Green trains can connect regions faster and cleaner. Hydrogen supply chains create new industries: tank and valve manufacturing, vehicle assembly, port refuelling facilities, and safety systems.

In this system, every kilometer of road, rail, or canal becomes an opportunity for innovation and employment. Young Africans can be trained as hydrogen technicians, safety officers, and system designers. Logistics becomes not just movement, but empowerment.

Global and Continental Impact

Africa's clean logistics corridors also redefine its position in the global economy. With hydrogen fuelling ships, trucks, and factories, the continent can export goods with near-zero emissions - a powerful competitive advantage in a world shifting toward green trade.

These corridors also make Africa a key player in decarbonized global shipping. Ports along the Red Sea, Indian Ocean, and Atlantic coasts can serve as **hydrogen bunkering hubs**, supplying vessels traveling between Asia, Europe, and the Americas. This positions Africa not at the edge of global trade, but at its clean energy center.

Logistics as a Symbol of Unity

Clean logistics connect more than products - they connect people and nations. When energy and goods flow smoothly, borders soften, cooperation grows, and prosperity spreads. A truck that runs on hydrogen from Ethiopia to Ghana does more than transport cargo; it carries the spirit of a united Africa.

In the future, when ships leave African ports powered by hydrogen instead of oil, they will symbolize more than progress. They will embody a deeper truth: that Africa has chosen a path of motion without destruction, connection without dependence, and growth without pollution.

The hydrogen corridors of Africa will carry not just fuel, but the dreams of a continent learning to move by its own light.

5.5 The African Grid: From Fragmentation to Flow

For decades, Africa's power systems have been defined by fragmentation. Each country built its own grid, each utility its own plan, and each project its own boundary. Cables stopped at borders, and so did progress. The result has been an energy map full of bright spots and dark gaps - a continent of potential that rarely connects.

The time has come to turn that fragmentation into **flow**. The **African Grid** must evolve from separate national systems into one living network - a continental circuit that allows energy to move as freely as ideas and trade.

From Islands to Interconnection

A connected grid is more than technology; it is trust. It means that one nation can rely on the next, that power can move where it is needed most. East Africa's hydropower can balance the Sahel's solar. The Congo's rivers can support Southern Africa's industry. North Africa's coastal hubs can export to Europe while feeding inland demand. When interconnected, Africa's resources no longer compete - they cooperate.

The **GERD+** system embodies this principle. It was never meant to serve Ethiopia alone. It was designed as an anchor for the regional grid - a steady source of clean power that can stabilize neighbours and feed into the broader **Pan-African Energy Network**. The same transmission lines that carry electricity from GERD can also link into hydrogen conversion stations, creating dual pathways for energy flow - electric and molecular.

In a connected Africa, no region stands idle while another struggles. The grid becomes a web of solidarity. When one river runs low, sunlight elsewhere fills the gap. When one system faces surplus, hydrogen production absorbs it. The flow continues - balanced, circular, and alive.

Technology and Design for Flow

Building a continental grid requires more than cables. It needs intelligence. Smart-grid technologies can track energy in real time, optimizing distribution, preventing waste, and ensuring stability across regions. These systems can automatically divert excess electricity to hydrogen plants or storage facilities, transforming volatility into value.

Cross-border transmission corridors - like those linking GERD+ to Sudan, Kenya, and the Pan-African Canal - will serve as **energy highways**, carrying both electrons and economic growth. Regional power pools, such as those already forming in East and Southern Africa, will evolve into interconnected hubs under a unified continental framework.

Over time, Africa's grid will function like a living body: hydropower as the heart, solar fields as the lungs, hydrogen as the bloodstream, and the people as the mind guiding it all.

From National Control to Collective Ownership

Energy integration requires a shift in mindset. It demands that nations move from isolated ownership to shared stewardship. Power must be seen not as a tool of control, but as a form of connection. The African Grid will not belong to one government or one corporation; it will belong to the people it serves - collectively, transparently, and sustainably.

This collective ownership also strengthens resilience. When infrastructure is shared, maintenance is shared. When planning is coordinated, resources are optimized. And when governance is continental, accountability is strengthened.

Economic and Human Impact

An integrated grid means more than efficiency - it means transformation. Electricity will reach industries, schools, and farms that have waited generations for power. Local businesses will thrive on stable energy prices. Cross-border trade will expand as supply chains stabilize. And most importantly, millions of homes will finally see the light.

A continental grid also fosters a new sense of unity. When one light flickers in one nation, others will help keep it on. Energy will no longer divide; it will unite. The grid becomes a living metaphor for Africa's destiny - connected, creative, and compassionate.

From Fragmentation to Flow

Flow is life. Nature never wastes motion. Rivers do not stop at borders, and sunlight does not recognize flags. Africa's grid must learn from the same principle. It must move as one current, guided by intelligence and harmony.

GERD+ is the beginning of that flow - the spark that shows how energy can move without loss, without conflict, and without end.

When Africa's grid becomes one, so will its destiny. Power will no longer be a privilege of a few but the pulse of an entire continent - steady, strong, and shared.

6.0 THE PAN-AFRICAN CANAL AND ENERGY CORRIDORS

Water has always shaped Africa's destiny. It carved valleys, fed civilizations, and connected people long before borders were drawn. Today, in the age of renewable energy, water is once again returning to its rightful place - not only as a source of life, but as a path of transformation. The **Pan-African Canal**, together with the **GERD+ energy system**, forms the backbone of that renewal: a network where trade, transport, irrigation, and clean energy move together in harmony.

The **Pan-African Canal Project**, envisioned by the author, is a continental infrastructure designed to unite landlocked nations with the sea and to restore balance to Africa's geography. It is not only a waterway; it is a symbol of connection - linking rivers, economies, and cultures. When integrated with GERD+, it becomes an **Energy Corridor** that does more than move goods; it moves power, knowledge, and cooperation.

The Vision of Flow

The canal is built on a simple truth: what flows survives. It carries ships, energy pipelines, irrigation systems, and floating solar fields along its route, forming a continuous band of productivity across the continent. Its waters nourish farms. Its banks host hydrogen plants and clean logistics zones. Its channels reduce transportation costs while opening trade to millions of people.

Through this design, Africa's inland nations - from Ethiopia and Uganda to Chad, Niger, and Malawi - gain access to global markets without losing sovereignty. The continent begins to function not as a collection of separated economies but as one continuous organism of flow.

Energy Corridors as Lifelines

Each section of the canal doubles as an **energy corridor**, carrying pipelines for hydrogen and ammonia alongside power cables and transport routes. These corridors connect inland energy production centers like **GERD+** with seaports on the Red Sea, Indian Ocean, and Atlantic.

Floating solar panels on the canal's surface generate electricity while reducing evaporation, while nearby electrolysis stations use that electricity to produce hydrogen. The same water that moves ships and irrigates farms also powers turbines and electrolysis units. Energy and water become one system - a perfect example of circular design.

These corridors are the **arteries of the African Energy Renaissance**. They do not only carry energy outward; they carry development inward. Industrial parks, research centers, and training institutes line the canal routes, turning infrastructure into education and employment. A fisherman in Ethiopia, a farmer in Sudan, and a shipbuilder in Mozambique all become part of the same continental story.

Trade, Transport, and Transformation

The Pan-African Canal will dramatically reduce transport costs across the continent. Goods that once travelled by slow and expensive land routes can move swiftly and sustainably by water. Hydrogen-powered barges and ships will replace diesel vessels, turning trade routes into zero-emission corridors. The canal becomes both highway and lifeline - carrying fertilizer upstream, food downstream, and ideas in every direction.

Energy corridors built alongside the canal also ensure that no section of the continent remains energy-poor. As pipelines and grids stretch across regions, power can be shared in times of shortage or excess. When one region faces drought, another's solar or hydrogen production balances the flow.

Peace Through Connection

The Pan-African Canal is as much a peace project as it is an engineering one. When nations share water and energy, they share destiny. The canal and its corridors create interdependence that replaces competition with cooperation. The flow of water becomes the flow of trust.

The African Union's Agenda 2063 calls for integrated infrastructure and sustainable development. The canal fulfills that vision in physical form - a waterway of unity, a renewable artery of prosperity.

A Living Corridor

The Pan-African Canal and its energy corridors embody the philosophy of a **living Africa**. Nothing is wasted. Every movement serves many purposes - irrigation, trade, energy, and employment. Water moves light; light feeds water. Together they sustain growth without harming the land.

This integration of water and power, agriculture and industry, is more than design - it is destiny. Africa's rivers have always sought the sea. Now, at last, their journey will complete the circle - not through colonial trade routes, but through a self-made, self-powered network of freedom.

The canal and GERD+ are not separate visions. They are two hands of the same body - one carrying the flow of water, the other the flow of energy. Together, they turn Africa's geography into a living system of light, food, and motion.

When complete, the Pan-African Canal will not just connect seas. It will connect futures - a pathway where water, sunlight, and humanity travel side by side toward an Africa that finally moves as one.

6.1 Water as Infrastructure for Energy

Water is more than a resource - it is the original infrastructure of life. Before there were roads or wires, water connected everything. It carved paths through stone, carried seeds across valleys, and gave birth to civilizations. For Africa, water has always been destiny. The Nile, the Congo, the Niger, the Zambezi - each one has written part of the continent's story. Now, in the age of clean energy, water will once again lead the way - not only as a source of power, but as the very **infrastructure for energy**.

In the **GERD+** and **Pan-African Canal** vision, water is both pathway and power. It moves ships, drives turbines, feeds farms, and carries hydrogen. It is the one element that can unite the continent's geography and technology. Every river, lake, and reservoir becomes part of a continental grid - a living web of flow and function.

From Natural Flow to Energy Network

Traditional infrastructure divides functions: roads for transport, pipes for water, cables for power. But water combines them all. It already flows across borders and landscapes, linking ecosystems and people. By treating water as integrated infrastructure, Africa can multiply its efficiency and value.

For example, the reservoir at **GERD+** does more than generate electricity. It supports floating solar systems that capture sunlight and shade the water, reducing evaporation. The canal that carries irrigation water can also transport ammonia for fertilizer and hydrogen for export. Pumping stations that lift water for agriculture can also serve as nodes for electricity generation. In this way, water becomes the skeleton of a renewable energy ecosystem.

Water-Powered Logistics

Waterways are the most natural form of clean transport. They require no pavement, consume little energy, and can move vast quantities of goods. When combined with **hydrogen-powered ships**, they become the backbone of sustainable trade. Every kilometer of canal and river can double as a green highway - carrying both water for crops and goods for markets, while using the same hydrogen that Africa produces through its clean energy systems.

The Water–Energy–Food Nexus

The integration of water and energy also secures agriculture. Hydropower and irrigation must no longer compete; they must collaborate. The electricity produced by dams can power pumps and fertilizer production, ensuring food self-sufficiency. In return, the food systems sustain the local economies that maintain and protect the energy infrastructure. This circular system closes the loop between survival and progress.

In the **GERD+–Pan-African Canal** model, this nexus is no longer a theory - it is a design. Water provides the structure. Energy provides the motion. Food provides the meaning. Each depends on the others.

The Wisdom of Flow

Water teaches balance. It does not rush where it is not welcome. It adapts to shape and slope, finding its way around obstacles until it reaches the sea. Energy systems must follow the same wisdom. When designed around water's rhythm, they become resilient - able to store in reservoirs, flow through pipelines, and renew through rainfall and sunlight.

By building with water, Africa avoids the mistake of building against nature. Every dam, canal, and reservoir can serve multiple generations if designed with care and humility. The goal is not to dominate rivers, but to cooperate with them - to use their flow as guidance, not resistance.

Continental Integration through Water

The Pan-African Canal extends this philosophy across the continent. It links rivers and lakes into one navigable, productive system - a blue network that supports trade, transport, and energy exchange. It gives landlocked nations access to the sea and coastal nations access to inland markets. More importantly, it reminds us that Africa's strength lies in connection, not isolation.

Water is both memory and promise. It remembers where it has flowed and teaches us how to move forward. It shows that unity is not a political idea but a natural law. Rivers do not stop at borders; they continue until they find purpose. Africa's energy must do the same.

In treating water as infrastructure for energy, Africa redefines development itself. It turns nature's pathways into engines of creation and transforms geography from challenge to advantage.

When water becomes the backbone of energy, every drop serves life twice - once as nourishment, and again as light.

6.2 The Pan-African Canal: The Continental Artery of Trade and Power

Africa has always been a continent of rivers - yet most of its great waters flow away from its landlocked heart. Sixteen nations sit without direct sea access, their trade and transport costs multiplied, their economies slowed by geography. The **Pan-African Canal** was born to change that - to turn isolation into integration, and distance into destiny.

The canal is not a single channel; it is a **continental artery**, a connected system of waterways, ports, and logistics hubs that bind the interior of Africa to its coasts. It transforms the map of Africa - drawing new lines of flow that mirror the paths of rivers and trade that existed long before modern borders.

The Vision of Connection

At its essence, the canal is a message: Africa must move together. It carries ships, energy, water, and opportunity through regions once considered unreachable. From the Ethiopian highlands to the Central African plains, from the Sahel to the southern lakes, the canal opens routes for commerce, irrigation, and clean transport.

But this is more than engineering. It is **continental architecture** - the physical expression of Africa's unity. Where colonial powers once built borders, the canal builds bridges. Where deserts once divided, the canal unites.

Trade Through Water

Water transport remains the most efficient means of moving goods. One barge can carry the load of hundreds of trucks with a fraction of the energy.

By integrating with **hydrogen-powered ships**, the Pan-African Canal makes this ancient mode of movement both modern and clean. It allows Africa to move raw materials, food, and finished goods at minimal cost and with zero emissions.

Ports along the canal will serve as **energy and logistics hubs**, powered by the same renewable electricity that drives GERD+ and its sister projects. Fertilizers from green ammonia plants, manufactured goods, and agricultural produce can travel across borders quickly and cheaply.

Each terminal becomes more than a port - it becomes a **living market**, a place where water, trade, and innovation meet.

Power Through Flow

The canal is also an **energy system**. Floating solar fields on its surface generate electricity that powers pumps, locks, and navigation. Pipelines for hydrogen and ammonia run along its banks, carrying energy produced inland toward the coasts. The same flow that carries ships carries power - water and energy intertwined, moving as one.

By integrating with **GERD+**, the canal receives a steady supply of clean energy while offering a transportation network for Africa's hydrogen exports. This connection forms a **Pan-African Energy Corridor**, turning the continent's waterways into pathways of prosperity.

Irrigation, Industry, and Inclusion

Along its length, the canal will feed irrigation systems that transform drylands into farmland. The water will sustain forests, crops, and fisheries while supporting eco-industrial zones. Farmers, traders, and engineers will all benefit from the same infrastructure - one channel serving many purposes.

The canal also ensures that industrialization is **inclusive**. Small ports and regional distribution centers will give local producers access to continental markets. Rural communities that once stood far from national highways will now stand beside Africa's new inland sea.

A Continental Artery of Peace

The Pan-African Canal is a peace project disguised as infrastructure. When nations share water, energy, and trade routes, they share interests and stability. The canal binds regions together through cooperation, replacing the politics of competition with the logic of flow.

It will also stand as a visible expression of African unity - a new Nile of the future, flowing not for one nation, but for all.

Built for the Living Century

The canal is designed not for the past but for the living century ahead. It aligns with Africa's climate and geography, powered by hydrogen, protected by renewable systems, and adaptable to rising technologies. It is not an imitation of foreign megaprojects but an **original African design** - shaped by African landscapes and created for African needs.

When complete, the Pan-African Canal will be the **circulatory system** of the continent's new economy. It will carry trade, power, and hope from coast to coast.

A continent that once waited for ships at foreign ports will now send its own - built, powered, and guided by its own light.

6.3 Green Hydrogen Logistics and Maritime Export Systems

The world is entering an age when energy no longer flows in barrels but in molecules of light. Green hydrogen and ammonia are becoming the new currencies of global trade - clean, transportable, and essential for the decarbonized industries of the future. In this new order, Africa stands not at the periphery but at the center. The **GERD+ system**, together with the **Pan-African Canal**, gives the continent both the power to produce and the pathways to deliver.

Hydrogen as a Global Export

Hydrogen produced from Africa's rivers and sunlight will be among the cleanest in the world. Using renewable hydropower and floating solar, the GERD+ complex can produce millions of tons of green hydrogen and ammonia each year. These fuels can be shipped to global markets where demand is rising - in Europe, Asia, and beyond - for use in shipping, steelmaking, and power generation.

Hydrogen exports are not about selling raw energy. They are about **selling transformation** - molecules that carry Africa's ingenuity, its sustainability, and its sovereignty. Every shipment of clean hydrogen or ammonia represents a contribution to the world's climate solution, created entirely by African innovation.

The Role of the Pan-African Canal

The **Pan-African Canal** provides the arteries for this global connection. Along its course, pipelines transport hydrogen and ammonia from inland production centers like GERD+ and the Niger or Congo Hubs to coastal terminals. These terminals become **Green Maritime Ports** - ports that export energy and goods without carbon emissions.

Hydrogen-powered barges travel along the canal, transporting energy as fuel while being fuelled by the same energy they carry - a perfect circle of sustainability. Alongside, floating solar panels generate power for port operations, irrigation, and local communities.

This system eliminates the need for long-distance diesel transport or heavy trucking. Water becomes both highway and powerline, carrying clean energy across the continent and out into the world.

Maritime Export Systems

At Africa's coasts, the **Green Hydrogen Maritime Export System** connects the continent to global trade networks. Terminals along the **Red Sea, the Gulf of Aden, the Indian Ocean, and the Atlantic** are equipped with liquefaction, conversion, and refuelling facilities for hydrogen and ammonia.

From these ports, Africa can supply hydrogen fuel for international shipping - a market that is rapidly transforming under new global regulations for zero-emission transport. African ports like Djibouti, Lamu, Beira, and Walvis Bay can become **global refuelling hubs**, serving vessels traveling between Asia, Europe, and the Americas.

Hydrogen-based fuels such as **green ammonia** will power large ocean vessels, while compressed or liquefied hydrogen will serve regional routes. African-built ships, powered by African energy, will sail the same waters once dominated by imported fossil fuels - this time carrying a cleaner story.

Circular Logistics and Local Benefits

Green hydrogen logistics are not just about exports; they also serve domestic industries. Pipelines and ports along the canal will supply hydrogen to fertilizer plants, clean transport systems, and local grids. Ammonia from GERD+ can be distributed to farmers along the route, while hydrogen refuelling stations support buses and trains in regional cities.

This **circular logistics system** ensures that every export corridor also feeds local development. The same ship that leaves port carrying hydrogen fuel may return carrying goods manufactured in African industries powered by that very fuel. Energy becomes both export and enabler - creating value twice.

A New Maritime Civilization

Through green hydrogen and the canal system, Africa can build what may be called a **Maritime Civilization of Light** - a trade network powered entirely by renewable energy, connecting oceans, people, and industries. This civilization is not about domination or extraction but about cooperation, balance, and shared progress.

Africa's ports will no longer stand as waiting points for imports but as launching points for innovation. Each vessel that departs from these ports will carry a message: that Africa has chosen to lead through creation, not consumption.

Global and Continental Impact

As global demand for hydrogen grows, Africa's exports will generate new income, strengthen currencies, and create millions of jobs. But the deeper impact is symbolic. It will redefine how the world perceives Africa - not as a source of raw materials, but as a **source of clean power and wisdom**.

The ships powered by hydrogen from GERD+ will not only move across the sea; they will carry Africa's light into the future.

When hydrogen flows from river to port and from port to planet, the circle of transformation will be complete. The continent that once fuelled others' revolutions will finally power its own - quietly, cleanly, and endlessly.

6.4 Continental Irrigation and Food Security

A continent cannot be strong if it cannot feed itself. Energy without food creates imbalance; food without energy limits progress. True transformation must unite the two. The **GERD+** and **Pan-African Canal** systems are designed to do exactly that - to bring together power, water, and agriculture into one living network of prosperity.

Africa's farmlands are vast, but much of them depend on unreliable rainfall. The continent holds about 60 percent of the world's uncultivated arable land, yet millions still face hunger and food insecurity. This is not a problem of soil - it is a problem of water and planning. With the right irrigation systems powered by renewable energy, Africa can become not just food-secure but food-abundant.

Energy-Driven Irrigation

The GERD+ model makes irrigation fully sustainable. The hydropower generated at the dam and the floating solar panels above it can power vast networks of electric pumps, desalination plants, and smart irrigation systems. Instead of diesel-powered engines that pollute and cost foreign currency, farmers can draw water using clean electricity or hydrogen-based microturbines.

Hydrogen also plays a crucial role. Its byproduct - oxygen - supports water purification, and its energy can be used to run large-scale desalination projects along coasts and canals. Green ammonia produced at GERD+ becomes fertilizer, closing the loop between energy and agriculture. In this circular system, **water powers food, and food sustains life**.

The Pan-African Canal as a Food Corridor

The **Pan-African Canal** transforms not only transport but also the way Africa grows its food. Along its length, irrigation channels branch into farmlands, carrying water to areas that once lay barren. Each section of the canal becomes a green belt - a corridor of crops, forests, and fisheries.

Water storage and distribution are managed by renewable-powered control systems, ensuring fair and efficient use. Floating solar arrays on canal sections generate electricity for local farmers. The canal's banks host agro-industrial parks, food processing centers, and cold-chain facilities powered by hydrogen and solar. This integration drastically reduces post-harvest losses and increases exports.

In this design, every kilometer of the canal is a **field of opportunity** - producing energy, food, and employment all at once.

Hydrogen Fertilizer and Soil Renewal

Ammonia, created from hydrogen and nitrogen, is more than an export commodity - it is the foundation of food security. The GERD+ system ensures that Africa will no longer depend on imported fertilizers that strain economies and supply chains. Instead, locally produced green ammonia can fertilize African soil with zero emissions.

This shift not only boosts productivity but also revives soil health. Clean fertilizer, combined with sustainable irrigation, rebuilds degraded lands and reduces chemical runoff. Africa can grow more while preserving its ecosystems.

From Scarcity to Surplus

The integration of energy and irrigation can turn Africa from an importer of food into a global supplier. Vast drylands in the Sahel, East Africa, and Southern Africa can be reclaimed for cultivation using renewable-powered pumps and canal-fed systems. Regions once threatened by drought can become stable sources of food and income.

This system also promotes **regional food cooperation**. Countries connected by the Pan-African Canal can share both water and surplus produce. Grain from Ethiopia, fruits from Malawi, and vegetables from Sudan can travel quickly and cleanly across the continent. In times of crisis, this interconnected network ensures no region is left behind.

Food Security as Peace

Food security is peace. When people eat well, they live well. When nations can feed their citizens, they stabilize. The GERD+ and Pan-African Canal initiatives embody this truth. They turn energy into nourishment, infrastructure into inclusion, and water into wellbeing.

Africa's irrigation revolution is not about endless expansion - it is about balance. It is about growing wisely, preserving soil, and aligning agriculture with climate. It is about restoring the dignity of feeding ourselves with our own hands, our own energy, and our own knowledge.

A Living Circle of Life

The circle is simple yet powerful:

Water flows from the dam → Energy powers irrigation → Crops grow → Waste returns as biomass → Hydrogen and ammonia fuel the next cycle.

Nothing is lost; everything lives again.

When Africa unites energy with food, the continent becomes self-sufficient, strong, and compassionate. The fields that bloom under sunlight and the water that flows through canals will not just feed stomachs - they will feed confidence, peace, and pride.

The GERD+ system shows that the power to feed the world begins with the power to feed ourselves.

6.5 How the Canal and GERD+ Reinforce Each Other

The **Pan-African Canal** and **GERD+** are two ideas born from the same philosophy - that Africa's strength lies in connection, not separation. Each project has its own purpose, but together, they create something larger than both: a **continental system of balance**, where water and energy flow in harmony, and every movement serves multiple goals.

A Partnership Between Water and Light

GERD+ harnesses water to produce clean power. The Pan-African Canal carries water to feed people and move trade. When they are integrated, energy and water become partners - each amplifying the other's purpose. The dam produces the power that runs the canal's locks, pumps, irrigation systems, and logistics hubs. In return, the canal provides water management and transport routes that strengthen GERD+ operations.

The floating solar fields that cover sections of the reservoir and canal not only generate electricity but also reduce evaporation, preserving the water that sustains both systems. Every watt of energy and every drop of water reinforce the same purpose: renewal.

Hydrogen and Water in Motion

Hydrogen, the heart of GERD+, flows naturally along the canal's path. Pipelines running beside or within the canal's infrastructure carry hydrogen and ammonia from inland production centers to coastal terminals. As ships and barges travel the canal, they can refuel with clean hydrogen - the same energy born from the dam's electricity.

In this way, the canal becomes more than a trade route; it becomes an **energy river**, carrying both goods and power. It turns logistics into sustainability, ensuring that every vessel moving through Africa's heart is powered by African light and water.

Irrigation and Power for Food and Peace

The canal irrigates farmland across multiple countries. GERD+ provides the energy to pump and distribute that water. Together, they create a self-sufficient system of food and energy security. This dual functionality - electricity for pumps, water for fields - ensures that no part of the continent suffers from drought or hunger while others thrive.

By linking irrigation and power, the system stabilizes communities, reduces migration pressures, and transforms regions once vulnerable to climate shocks into green zones of productivity. Water and light, united, become instruments of peace.

Trade and Energy Synergy

The canal gives GERD+ a highway to the world. Hydrogen and ammonia produced inland can be transported to coastal ports without reliance on carbon-heavy roads. This drastically lowers transport costs and emissions, making African clean energy competitive on global markets.

In return, the revenue from hydrogen exports supports further canal development, port upgrades, and irrigation expansion. The cycle strengthens itself: **energy builds trade, trade funds energy**.

Shared Infrastructure, Shared Ownership

Both projects are rooted in the same principle - **African ownership of African transformation**. GERD+ shows that Africa can build world-class infrastructure through its own determination. The Pan-African Canal extends that model to a continental scale, promoting cooperative development where nations share benefits, maintenance, and governance.

The canal's governance framework, built around the African Union and regional communities, mirrors the inclusive philosophy of GERD+: no one dominates, and all benefit. This balance ensures that cooperation, not competition, becomes the default mode of progress.

Environmental Harmony

The union of GERD+ and the canal reflects ecological intelligence. Together, they manage water sustainably, balance ecosystems, and reduce carbon emissions across transport and agriculture. The energy that drives the canal is clean; the water that cools the land is conserved.

Floating solar panels, reservoir management systems, and hydrogen-fuelled vessels minimize environmental impact while maximizing efficiency. This is not industrialization against nature - it is **civilization with nature**.

A Unified Continental Symbol

When seen together, GERD+ and the Pan-African Canal form a single living image: a river of light flowing through a river of water. One represents Africa's capacity to create, the other its capacity to connect.

Their union is symbolic of what the continent itself must become - diverse yet united, self-powered yet interdependent, modern yet harmonious with the natural world.

In every sense, GERD+ and the canal reinforce one another. The dam empowers the canal; the canal extends the dam's reach. One gives energy, the other gives motion. Together, they give meaning.

They stand as proof that Africa's future will not come from separating systems, but from integrating them - from seeing that water, energy, food, and people are all part of one living design.

When the waters of the canal carry ships powered by hydrogen from GERD+, the circle of renewal will be complete: **Africa, moving by its own light and flowing by its own water.**

7.0 THE HUMAN SIDE OF ENERGY

Every great system begins and ends with people. Dams, canals, and pipelines may carry water and power, but it is human beings who give them purpose. Without people, there is no progress; without values, there is no direction. The success of **GERD+** and the **Pan-African Canal** will not be measured only in megawatts or kilometers - it will be measured in how deeply they transform lives.

Africa's energy renaissance is not just a technical project; it is a **human story**. It is about restoring dignity, creating opportunity, and giving every young African the tools to shape their own destiny. Energy is not only electricity - it is empowerment. It gives birth to education, industry, health, and confidence.

Energy as Human Freedom

When a community receives electricity for the first time, it receives more than light. It receives freedom - the freedom to learn after sunset, to refrigerate medicine, to communicate, to innovate, to dream. Energy breaks the cycle of limitation. It transforms survival into creativity.

GERD+ embodies this transformation. By expanding renewable power across borders, it gives people more than power supply; it gives them **power of choice**. Rural families no longer depend on firewood or expensive fuel. Students can study at night. Farmers can irrigate their land efficiently. Small industries can grow without pollution. Energy, in human terms, is independence.

Jobs, Skills, and Knowledge

The GERD+ system and the Pan-African Canal will create millions of jobs - not just during construction, but for generations afterward. Engineers, electricians, welders, architects, software developers, port workers, and hydrogen technicians will all find opportunity in this new energy ecosystem.

But beyond employment, these projects build **human capacity**. They establish training centers, research institutions, and innovation hubs along the canal and energy corridors. Young Africans will learn to design, maintain, and improve renewable systems using their own hands and minds. Energy development becomes education in motion.

Every new skill learned becomes a light turned on in another life. Every trained technician becomes a multiplier of knowledge, teaching others. This is how real development spreads - not by donation, but by participation.

Energy and Gender Equality

Clean energy also reshapes social life, particularly for women and girls. In many rural areas, women spend hours gathering firewood or fetching water. Renewable electricity and water access reduce that burden, freeing time for education, business, and family. The shift from manual survival to productive opportunity empowers women to lead in every sphere of life - agriculture, enterprise, governance, and science.

The GERD+ initiative places gender inclusion at its core. Women will not only benefit from energy access; they will be trained, employed, and represented in every stage of design and management. Equality is not a goal to achieve later - it is part of the structure being built now.

Community and Cooperation

Energy systems can unite or divide. If built with exclusion, they create inequality; if built with inclusion, they create unity. GERD+ and the Pan-African Canal are designed as **shared systems** - owned collectively, serving equally.

Community participation in planning and benefit-sharing ensures that development does not displace but empowers. Local people help manage water, protect ecosystems, and operate small-scale grids and irrigation projects. The energy that lights their homes also strengthens their voice.

Culture, Identity, and Pride

The human side of energy is also cultural. When Africa builds its own systems - designed by African minds, constructed by African hands - it creates not just infrastructure but identity. GERD+ is a monument to the spirit of self-reliance. The Pan-African Canal is a statement of unity. Together they tell a story that every African child can inherit with pride: *We did this ourselves.*

Clean energy is not only about climate or technology. It is about restoring confidence - showing that progress can be born from our rivers, our sunlight, our wisdom.

Energy as Peace

Perhaps the most powerful human outcome of renewable energy is peace. When nations share power grids, water routes, and economic benefits, conflict loses its purpose. Energy interdependence builds stability faster than diplomacy alone.

The GERD+ and Pan-African Canal systems encourage cooperation through shared prosperity. The light that flows from one country to another becomes a bridge, not a boundary.

A Living Energy Civilization

In the end, the purpose of GERD+ is not to produce electricity or hydrogen; it is to produce **life** - a way of living where power serves people and nature equally. This is what makes it human.

The real success of Africa's Energy Renaissance will not be seen in its machinery, but in its people - educated, employed, inspired, and united.

Energy is human potential set in motion. And when that potential moves, no force on Earth can stop it.

7.1 Energy for Education and Industry

Education and industry are the twin engines of transformation. One shapes minds; the other shapes materials. Together, they build civilizations. But neither can grow in the dark. Without reliable and affordable energy, schools stay closed after sunset, machines fall silent, and dreams remain idle. The **GERD+** system was designed to change this - to make energy not only a source of power, but a source of progress.

Lighting the Mind

A single light bulb in a classroom can change the course of a child's life. Electricity means study after dark, access to technology, and connection to the world. In too many parts of Africa, education still stops when the sun sets. GERD+ turns daylight into nightlight - enabling continuous learning powered by clean energy.

Schools along the **Pan-African Canal** and in rural communities connected to regional grids will receive stable electricity from renewable sources. Floating solar stations and mini-hydropower systems ensure that even isolated schools never lose power. This creates a new rhythm of education - one not bound by the rising and setting of the sun.

With internet access powered by renewable energy, students can learn science, engineering, and coding. Digital classrooms will link children in Ethiopia, Kenya, Niger, and Ghana - forming a **continental classroom** connected by light.

Universities as Energy Hubs

Africa's universities will play a central role in this system. The GERD+ and Pan-African Canal projects will establish **centers of excellence in renewable energy and hydrogen technology**. These institutions will train engineers, researchers, and technicians to design, operate, and improve Africa's own systems.

Instead of importing expertise, Africa will export it. The same students who learn how to build hydrogen electrolyzers or maintain floating solar fields will later teach others and establish their own enterprises. Energy education becomes an unbroken chain of empowerment - knowledge multiplying like sunlight.

Each campus can serve as a **microgrid laboratory**, powered entirely by renewable sources and connected to nearby communities. Universities will not just teach energy - they will live it.

Energy for Industry

Industry is the practical extension of education. What the mind imagines, the hand builds. GERD+ gives those hands the power they need. Stable, affordable energy means factories that run efficiently, food that is processed locally, and materials that are transformed rather than exported in raw form.

Africa's industrialization has long been limited by energy shortages and high costs. Green hydrogen and ammonia from GERD+ remove these barriers. They provide the fuel for steelmaking, cement production, transport, and chemical industries - all powered by clean molecules rather than imported oil or coal.

Industrial parks along the **Pan-African Canal** will draw power directly from renewable grids and hydrogen pipelines. Local entrepreneurs can establish manufacturing plants, recycling facilities, and logistics centers without the burden of high fuel prices.

This new industrial model will not repeat the mistakes of the past. It will be **circular, clean, and community-based** - producing jobs without pollution, and growth without waste.

The Learning–Making Cycle

Education feeds industry, and industry feeds education. As universities train skilled workers, industries provide the resources to fund further research and innovation. This creates a **learning–making cycle** - a continuous loop where knowledge and work grow together.

Imagine a young African studying renewable energy in a campus powered by GERD+, then joining a hydrogen plant along the canal, then later founding a company that builds green engines. That is the rhythm of development - self-taught, self-powered, self-renewing.

Energy as a Teacher

Energy itself is a teacher. It shows us how systems connect, how nothing stands alone. In the same way, education and industry must be designed not as separate institutions, but as parts of one living ecosystem. Schools teach theory; industries teach application; together they teach wisdom.

The **human side of GERD+** is not just employment - it is enlightenment. The electricity flowing from the dam does more than spin machines; it illuminates minds.

When Africa's schools are powered, its industries will follow. And when both move together, the continent will not just rise - it will rise with grace, intelligence, and purpose.

7.2 Empowering Youth through New Skills

Africa's greatest resource is not its rivers, minerals, or sunlight. It is its people - especially its youth. More than sixty percent of Africans are under the age of twenty-five. This generation is the pulse of the continent, the living current that will determine whether Africa's future flows with purpose or remains trapped in potential. The **GERD+** system and **Pan-African Canal** are not only projects of infrastructure; they are classrooms of transformation, designed to empower this generation through new skills and new confidence.

A Generation Ready to Build

The youth of Africa do not lack talent; they lack platforms. They are born into an age of technology, connected through mobile networks, yet disconnected from opportunities to apply what they know. The renewable energy revolution offers them that platform. Every turbine, every solar field, every hydrogen plant is an open invitation for African hands and minds to shape the future.

Through GERD+, the continent can train hundreds of thousands of technicians, engineers, and energy managers in fields like hydrogen production, electrolysis, floating solar design, ammonia conversion, and grid management. These are not skills for the few - they are the foundation of an entirely new economic era.

Young Africans will no longer wait for jobs to arrive; they will **build them**.

Learning by Doing

The education model behind GERD+ is not limited to classrooms. Training happens where life happens - at construction sites, power plants, research centers, ports, and farms. Apprenticeship and hands-on learning become part of national energy programs.

For example, along the **Pan-African Canal**, youth can learn irrigation management, hydrogen safety, or maintenance of solar systems while contributing to their community's development. In industrial parks powered by GERD+, young technicians can assemble and repair components, gaining practical experience that leads directly to employment.

This approach transforms energy projects into universities without walls - living schools that combine theory with creation.

Innovation and Entrepreneurship

Empowering youth also means unlocking creativity. With access to clean power and digital tools, young Africans can become **innovators and entrepreneurs**. Startups can design affordable solar kits, green mobility solutions, recycling technologies, and agricultural tools powered by hydrogen or electricity.

GERD+ will establish **Innovation Labs** along its network - spaces where youth can prototype new solutions, supported by mentors, grants, and technical resources. The same energy that powers industries will power invention.

Through collaboration across borders, African youth will begin to think continentally - sharing ideas and building partnerships that transcend nationality. Energy becomes a unifier not just of systems, but of minds.

From Skill to Dignity

Skills are more than economic assets; they are expressions of dignity. When a young engineer designs a hydrogen valve, or a farmer installs a solar pump, they are not only earning - they are **belonging**. They see their place in the continent's renewal, knowing that progress is no longer imported but created by their own generation.

This dignity changes everything. It replaces dependency with self-belief, and hesitation with action. The youth who master these new skills will not only build infrastructure - they will build a new African mindset.

The New Work Culture

The renewable era brings a new kind of work culture - one that values precision, collaboration, and creativity. It is less about hierarchy and more about harmony. It demands responsibility to both people and the planet. GERD+ and its related systems will cultivate this culture, shaping a workforce that is technically skilled, environmentally conscious, and ethically grounded.

A Generation of Guardians

Empowered youth are not only workers; they are guardians of the future. They will maintain Africa's living systems - the dams, canals, forests, and cities that together form a circular civilization. Their role is not simply to sustain what exists, but to imagine what's next.

Through education, skill, and purpose, Africa's youth will carry forward the philosophy behind GERD+ - that the continent's transformation must come from within.

The power flowing from turbines and solar fields will be matched by another kind of power - the will, intelligence, and courage of a generation that knows its worth.

When African youth rise with skill, Africa itself rises with strength. The light that shines from GERD+ will reflect in their eyes - steady, determined, and full of life.

7.3 The Rise of African Engineers and Technicians

Every civilization has its builders - those who turn dreams into stone, water into power, and ideas into living systems. For Africa, that generation of builders is rising now. They are the engineers, technicians, and innovators who will translate vision into structure and ensure that the continent's transformation is not temporary, but enduring. The **GERD+** and **Pan-African Canal** systems are not only blueprints for energy and trade; they are workshops for the rise of a new African technical civilization.

From Dependency to Mastery

For too long, Africa's greatest projects have been designed elsewhere and built by others. Foreign contractors arrived with technology, executed the work, and left with the knowledge. Africa inherited infrastructure but not the ability to sustain or replicate it. GERD+ marks a turning point - a declaration that **Africa will build its own systems, maintain them, and teach others to do the same**.

The Grand Ethiopian Renaissance Dam itself proved that African engineers can master projects of global scale. With GERD+, that confidence expands to a continental level. The same hands that built the dam can now build hydrogen plants, floating solar fields, and canal systems. Knowledge is no longer outsourced - it is cultivated, refined, and shared across borders.

The New Builders of the Continent

The rise of African engineers and technicians is more than professional - it is cultural. It restores dignity to work, precision to practice, and pride to progress. These builders will not be anonymous workers but recognized creators of Africa's future.

Each engineer becomes part of a living network - a continental guild of creators whose work shapes both infrastructure and identity.

Technical Institutions and Learning Networks

GERD+ will give birth to **technical universities, hydrogen academies, and innovation institutes** across Africa. These institutions will not only teach; they will design, prototype, and export technology. Collaborative programs between African nations will allow knowledge to circulate freely, ensuring that no country is left behind.

In partnership with the Pan-African Canal Project, mobile training centers can travel from region to region - floating universities on barges or canal vessels - bringing education to remote areas. A student in Chad, Malawi, or Mozambique will have the same access to learning as one in Addis Ababa or Nairobi.

This is how a continent builds not just systems, but skill ecosystems.

Women in Engineering

The rise of African engineers must include the rise of **African women engineers**. Their participation is not an option; it is essential. The renewable energy transition is a chance to reset norms - to ensure that women lead as designers, project managers, and innovators. GERD+ will reserve training and leadership programs for women in engineering and technology, making equality not a slogan but a standard.

When African girls see women building dams, managing energy plants, and leading scientific teams, they will understand that leadership has no gender.

Technology Ownership and Innovation

To rise fully, Africa must not only learn technology - it must **own it**. Patents, software, and production must increasingly originate from African labs and factories. Hydrogen systems, solar panels, and control equipment can be designed, assembled, and improved locally.

This ownership ensures that progress cannot be reversed. When the tools of development are made in Africa, the continent becomes the author of its own modernization.

Building a Culture of Precision and Purpose

Engineering is both science and discipline. It requires integrity, focus, and care. The rise of African engineers will also mean the rise of a new professional ethic - one that values quality over speed, collaboration over competition, and sustainability over short-term profit.

GERD+ represents this ethos. It is not just a structure, but a statement: **we build carefully because we build for centuries**.

A Continental Identity of Creation

In the coming decades, African engineers will design canals that reshape trade routes, energy systems that power industries, and smart cities that live in balance with nature. The world will look to Africa not as a recipient of aid, but as a laboratory of innovation.

Every bolt tightened, every turbine installed, every solar panel adjusted will echo the same message: Africa can build, Africa can maintain, Africa can lead.

The engineers and technicians of this new era are not just workers of progress - they are **guardians of light**, ensuring that what Africa builds will never fade.

When a young African steps onto a dam, a canal, or a hydrogen plant and says, "I helped build this," the future will have already arrived.

7.4 Local Production, Local Knowledge

Real transformation begins when production and knowledge live where people live. For generations, Africa's materials have travelled abroad to be refined, manufactured, and sold back at a higher price. This pattern drained both value and confidence. The **GERD+** and **Pan-African Canal** systems are designed to reverse that flow - to ensure that what Africa produces, Africa also processes, applies, and perfects.

Local production and local knowledge are the two pillars of genuine independence. One gives economic strength, the other gives intellectual sovereignty. Together, they create a cycle of empowerment that no external system can replace.

From Raw Materials to Finished Value

Africa's rivers, sunlight, and minerals should serve Africa first. The clean power generated by **GERD+** will supply energy for factories, agro-industries, and laboratories along the **Pan-African Canal Corridor**. Instead of exporting raw hydrogen or ammonia alone, Africa can export finished products - fertilizers, fuels, equipment, and even manufactured hydrogen technologies.

Each factory built along this corridor becomes part of a living supply chain:

- **Hydrogen from GERD+** powers machinery and transportation.
- **Ammonia** becomes fertilizer for local farmers.
- **Electricity** drives production lines for solar panels, water pumps, and green vehicles.
- **Education hubs** nearby train technicians and engineers to operate and innovate.

The result is value that circulates within the continent before it moves outward. Every ton of product exported carries both economic and intellectual weight.

Knowledge as Infrastructure

Knowledge is the most durable form of infrastructure. Machines wear out, but understanding endures. The GERD+ system integrates **learning**

centers and technical institutes into every major site - making knowledge a permanent component of production.

In these centers, local workers study everything from electrolysis and storage management to irrigation design and canal maintenance. Training happens in multiple languages, respecting Africa's cultural diversity while promoting shared technical standards.

This approach ensures that even the most advanced technology can be maintained locally. A breakdown no longer requires foreign consultants or imported spare parts. African experts diagnose, repair, and improve their own systems - reducing dependency and building pride.

Technology Adapted to the Land

Local knowledge also means local design. Africa's technologies must fit its geography and people, not the other way around. Floating solar panels are engineered to withstand tropical storms. Hydrogen pipelines are built to resist desert heat. Irrigation systems are designed to protect soil moisture in semi-arid zones.

By grounding technology in local wisdom, Africa avoids waste and strengthens sustainability. Farmers, fishermen, and communities contribute traditional insights - understanding water cycles, soil conditions, and seasonal patterns that modern science often overlooks. The result is **technology shaped by nature and culture together**.

Community Ownership and Cooperative Production

Local production is not only about factories; it is about participation. Communities along the canal and energy corridors will be encouraged to form **cooperatives** that manage mini-grids, irrigation systems, and waste-to-value facilities. Profits are shared locally. Decisions are made collectively. Energy and knowledge stay within the community that sustains them.

When a village owns its solar field or hydrogen pump, it gains more than income - it gains control. People protect what they own and take pride in what they manage. This sense of ownership is the true foundation of sustainable development.

The Power of Replication

Local production also means local replication. Once a technology is mastered in one region, it can be copied and improved in another. Designs, blueprints, and processes will be shared openly across African nations through a **Pan-African Knowledge Network** supported by digital platforms and training programs.

A hydrogen compressor built in Ethiopia can be reproduced in Ghana. A floating solar system tested on the Nile can be installed on the Niger. Knowledge moves faster than cargo, and its movement costs nothing but cooperation.

Economic Freedom through Intelligence

When production and knowledge remain local, economies become self-sufficient. Imports decline, exports rise, and currencies stabilize. More importantly, **dignity returns** - the understanding that Africa's progress depends not on what it receives, but on what it creates.

The GERD+ model demonstrates that intelligence is the truest form of wealth. Once the knowledge to build, repair, and innovate is widely shared, no resource can be taken away and no dream delayed.

Local production and local knowledge are not just strategies; they are philosophies. They affirm that Africa's solutions must grow from African soil, African sunlight, and African minds.

When a child in a village sees clean electricity, locally made tools, and trained neighbours who maintain them, they will know - without being told - that Africa's future is already here, and it was built right where they stand.

7.5 Energy as a Path to Dignity

Energy is more than electricity. It is light for the mind, warmth for the home, and movement for the dream. It gives rhythm to daily life and confidence to nations. When people have reliable, clean, and affordable energy, they can live with dignity - not in dependence, but in purpose.

The story of **GERD+** and the **Pan-African Canal** is, at its core, a story of dignity restored. It is about turning power from something Africa receives into something Africa creates. It is about replacing limitation with possibility.

The Meaning of Dignity

Dignity begins with self-sufficiency. A person who can light their home, water their field, and feed their family does not wait for permission to live. A continent that can power its industries and sustain its people does not need to ask for help to grow. Dignity is the quiet strength that comes from knowing - *we can do it ourselves.*

The GERD+ system brings that strength to life. It transforms the continent's natural gifts - water, sunlight, and intelligence - into tangible wellbeing. Electricity flows to villages that once knew only darkness. Hydrogen fuels industries that once stood still. Ammonia feeds soils that once went dry. Every watt of energy becomes a message: Africa can stand on its own.

The Human Light

When the first lightbulb turns on in a rural home, something invisible happens. Children study longer. Families gather safely. Businesses open after dusk. A sense of possibility spreads quietly through the air. This is the true illumination - not just the glow of bulbs, but the awakening of human confidence.

Clean energy removes the smoke from kitchens, the noise from generators, and the weight of fuel scarcity. It replaces exhaustion with health, frustration with progress. A woman who no longer spends hours gathering firewood gains time to learn, to lead, to build. That is dignity made visible.

Work with Meaning

Energy-powered work is not just employment - it is participation in creation. When a young technician installs a solar field, or an engineer designs a hydrogen plant, they are shaping the destiny of their people. They see the direct result of their labour in the light that reaches homes and the water that nourishes crops. Work connected to purpose dignifies both the worker and the work.

GERD+ will inspire thousands of such careers - rooted not in survival but in contribution. Every job in energy production, maintenance, research, or logistics will be a step toward collective pride.

Equality Through Access

Dignity cannot exist where inequality persists. Access to energy must be universal - not limited to cities or elites. The Pan-African Canal and its surrounding energy corridors are designed to spread opportunity, not concentrate it. By powering small villages, schools, farms, and industries equally, the system ensures that development moves with justice.

In this way, energy becomes a social equalizer. It breaks the invisible divide between those who have and those who have not. It brings the same light to every hand willing to build, every mind willing to learn.

Energy and the Spirit of Independence

Dignity is not pride without reason; it is confidence with evidence. When Africa generates, stores, and exports its own clean power, it gains a new identity - not the world's energy consumer, but its energy teacher. The technologies that emerge from GERD+, the hydrogen plants, the circular economy hubs, the canal's floating solar fields - these are more than facilities. They are proofs of ability.

Each success erases centuries of doubt. Each completed system writes a new definition of independence - one built not on isolation, but on self-reliant cooperation.

Harmony with Nature

True dignity respects the world that sustains us. GERD+ brings progress without pollution. It demonstrates that we can grow strong without harming the land, that development can flow gently like the river itself. This harmony restores Africa's spiritual balance - a reminder that human advancement and ecological health are not opposites, but allies.

When people see that their prosperity no longer costs the planet, they gain a deeper kind of pride: the pride of right action.

The Light Within

The greatest energy is not in turbines or hydrogen molecules. It is in people - their curiosity, discipline, and compassion. The purpose of energy is to unlock that inner light. GERD+ is therefore not a monument to machinery, but a reflection of what Africans can achieve when they believe in themselves.

Dignity comes when every person feels included in that achievement. When farmers, engineers, teachers, and children all recognize their role in the same living system, Africa becomes not just a continent of nations, but a continent of meaning.

The light that GERD+ spreads across rivers and cities will not only brighten homes - it will brighten hearts.

For in the end, energy is not about power over things; it is about **freedom to become**. And that freedom, once awakened, will never fade.

8.0 CIRCULAR AFRICA: WASTE TO VALUE

Nothing in nature is wasted. The leaf that falls becomes soil, the soil becomes food, and life begins again. This is the principle of the **circular system** - the wisdom that creation and renewal are one. For Africa to grow sustainably, it must rediscover this natural rhythm and make it the foundation of its economy. That is the spirit behind **Circular Africa: Waste to Value** - a vision where everything Africa produces is reused, regenerated, and reimagined for life.

The **GERD+** and **Pan-African Canal** systems are not just about producing energy; they are about producing *continuity*. Energy, water, and resources flow in loops, not lines. This same principle must guide Africa's industries, cities, and agriculture. Waste must no longer be an end - it must be a beginning.

The Circle as a Way of Life

Circular Africa begins with a change in thinking. Instead of extracting, using, and discarding, we design processes that renew themselves. We see waste as material waiting for a second purpose. Organic waste becomes fertilizer. Industrial waste becomes raw material. Energy waste becomes heat for new production.

The circular model mirrors nature itself. Every ecosystem recycles - forests, rivers, and even the human body. Africa's future industries must follow that same law of balance.

From Waste to Energy

Energy is the heartbeat of the circular system. Waste that once polluted cities and landscapes can now produce power. Through **biogas plants, waste-to-hydrogen conversion**, and **circular fuel systems**, organic matter can be transformed into clean energy.

Municipal waste in growing cities can feed anaerobic digesters that generate electricity. Agricultural residues can be processed into biohydrogen or used to fuel small turbines. Even wastewater can power treatment plants through microbial fuel cells. Each city becomes an energy source, not a burden.

In the **GERD+ ecosystem**, these waste-to-value systems can be connected to the main grid or used to power local mini-networks. Hydrogen derived from organic waste can blend with hydropower-based hydrogen to form a completely renewable energy mix - truly circular and clean.

Circular Agriculture

The Pan-African Canal will irrigate millions of hectares of farmland. But agriculture, too, must become circular. Crop residues can be converted into compost or biogas. Fertilizers produced from green ammonia at GERD+ can be recovered from soil through integrated nutrient cycles. Nothing leaves the system permanently - everything returns as value.

Fish farms, fruit plantations, and agroforestry projects along the canal can exchange resources: water enriched by one process nourishes another. This symbiosis mirrors Africa's original ecological wisdom - the understanding that every living thing sustains another.

Circular Cities and Industries

Africa's growing cities are often seen as challenges, but in a circular system, they become laboratories of regeneration. Urban planning guided by clean energy, recycling, and green architecture can turn waste into wealth.

Hydrogen-powered waste collection vehicles can transport materials to recycling hubs. Construction waste can be processed into new building materials, such as green bricks or composite panels. Factories can share waste heat through district energy systems, lowering costs and emissions.

The circular city model ensures that urban growth no longer means urban damage. Every building becomes part of the ecosystem - generating energy, reusing water, and reducing waste.

The Power of the Circular Mind

Technology alone will not create a circular Africa. What will make it possible is a change of mindset - a return to the traditional African understanding that nothing truly belongs to us; it all belongs to life. The continent has always lived close to this truth. In villages and forests, in crafts and cultures, Africans have long practiced reuse and regeneration. The circular economy is not a new idea for Africa; it is a modern expression of an ancient wisdom.

By connecting that wisdom with modern clean energy systems like GERD+, Africa can create an economy that grows without destruction - an economy of stewardship, not consumption.

From Pollution to Prosperity

Circular Africa also heals. Waste is not just an environmental issue; it is a social one. Poor waste management pollutes water, spreads disease, and diminishes dignity. Turning waste into energy and value cleans not only the land but the lives of those who live on it. It transforms pollution into employment, and neglect into pride.

Each recycling facility, waste-to-energy plant, and circular factory will become a symbol of renewal - proof that Africa's strength lies in transformation.

The Eternal Circle

When energy, water, agriculture, and industry are all linked in a circular system, nothing is truly lost. Light becomes movement, movement becomes growth, and growth becomes renewal. This is how a living continent sustains itself.

The circular economy is not just a model - it is a philosophy of respect. Respect for what we take, what we use, and what we return.

Circular Africa will prove to the world that sustainability is not a luxury; it is the natural way of life. The waste of the past will become the wealth of the future, and the cycle of loss will become a circle of abundance.

8.1 Turning Waste into Energy

In nature, waste does not exist. Every fallen leaf becomes soil, every current of air carries new life. It is only in human systems that waste appears - a sign that something has been designed without completion. The vision of **Circular Africa** restores that balance. Through the **GERD+** framework, even waste becomes energy, and energy becomes renewal.

Africa's cities, farms, and industries generate vast amounts of organic and solid waste every day. Instead of burying or burning it, this material can be transformed into power, fuel, and fertilizer. When energy systems learn to feed on what society discards, nothing is lost. What once polluted rivers and air now powers homes, farms, and schools.

From Refuse to Resource

Every city produces waste streams - food scraps, agricultural residues, paper, sewage, and biomass from forests. Through technologies such as **anaerobic digestion**, **gasification**, and **biohydrogen conversion**, this waste can be turned into usable energy forms.

- **Biogas plants** convert organic waste into methane for cooking or electricity generation.
- **Biohydrogen systems** use microbes or thermal processes to produce hydrogen from organic materials.
- **Waste heat recovery systems** capture unused industrial heat to generate power.
- **Wastewater treatment plants** use microbial fuel cells to produce electricity while cleaning water.

Each process fits naturally within the GERD+ ecosystem. The power generated from waste feeds the grid, complements hydropower and solar, and strengthens local resilience.

City Networks for Clean Power

Imagine every major African city as a **circular energy hub**. Waste collected from homes, markets, and industries is separated, processed, and converted into clean energy locally. Hydrogen refuelling stations can be powered by waste-derived biohydrogen, and electricity from biogas plants can sustain small industries.

In this system, trash trucks become energy carriers. They collect resources, not rubbish. Landfills disappear, replaced by recycling parks and waste-to-value centers. Cities become cleaner, healthier, and self-sufficient.

Agricultural Waste as Power

Across Africa's farmlands, millions of tons of crop residues and animal waste are left unused. These can provide large quantities of energy if properly managed. Straw, husks, and stalks can be turned into biogas, while livestock manure can power village-level digesters.

This approach gives farmers an extra source of income and reduces the need for firewood or imported fuel. The same energy that powers irrigation pumps and fertilizer production can also be created by the farms themselves - **agriculture feeding energy, and energy feeding agriculture**.

Hydrogen from Waste

In the GERD+ concept, hydrogen is not only produced from water using electricity. It can also be generated from organic waste through processes like **biomass gasification** and **dark fermentation**. This "biohydrogen" joins the clean hydrogen produced at GERD+ to form a truly diverse and circular energy system.

The oxygen released during these processes can purify water or enrich aquaculture systems along the **Pan-African Canal**, creating a complete environmental cycle. Nothing is wasted, not even air.

Social and Environmental Impact

Turning waste into energy is not merely an engineering task - it is an act of healing. It cleans cities, protects rivers, and creates jobs. The workers who once scavenged through landfills can become trained operators of clean facilities. Communities once living among pollution can become guardians of energy.

Each ton of waste processed represents dignity restored, health regained, and potential unlocked.

Integration into the GERD+ System

GERD+ acts as the **energy coordinator** of this circular model. Its stable hydropower ensures the reliability of city-level systems, while surplus power from waste can be fed back into regional grids. Floating solar systems can supply electricity to waste plants near the canal, ensuring continuous operation.

Through this integration, every layer of the energy cycle supports another. Waste becomes fuel, fuel becomes work, work becomes value.

A New Definition of Clean

Clean energy is not only about zero emissions - it is about zero neglect. It means caring for what is left behind and turning it into something useful. When waste becomes part of the design, cleanliness becomes continuous.

Africa can become the first continent to show that **progress and purity can coexist** - that cities can grow without garbage, and that power can rise from what was once thrown away.

When light shines from waste, it proves that nothing in life is useless - only unimagined.

8.2 The Circular Energy System

In a living continent, energy must behave like life itself - flowing, renewing, and returning. The **Circular Energy System** is the nervous system of a sustainable Africa. It connects production, use, and renewal in one continuous rhythm where nothing ends and nothing is wasted. This system, envisioned through **GERD+** and the **Pan-African Canal**, transforms how energy is generated, stored, and reused - making every kilowatt part of a greater cycle of regeneration.

From Linear to Circular

Most of the world still operates on a linear model: take, use, discard. Fossil fuels are extracted, burned once, and lost to the atmosphere. This approach feeds growth for a moment and emptiness for centuries. Africa can choose differently. The **circular energy model** begins where the linear model ends - by ensuring that every process creates value for another.

In a circular system:

- **Hydropower** and **solar energy** produce electricity.
- Surplus electricity produces **hydrogen** through electrolysis.
- Hydrogen combines with nitrogen to create **ammonia** for fertilizer and clean fuel.
- Agricultural waste and organic residues are turned into **biogas or biohydrogen**.
- Waste heat from industry powers nearby homes and schools.
- Oxygen released from hydrogen production purifies water and sustains fish farms.

Every action feeds another, forming a loop of life powered by intelligence rather than extraction.

GERD+ as the Heart of the Circle

At the center of this circular system lies **GERD+**, acting as the heart that keeps energy in motion. Hydropower flows steadily, while floating solar arrays pulse with daily rhythm. During times of excess, electricity generates hydrogen stored in the **Afar salt caverns**; during scarcity, that hydrogen returns as fuel.

Nothing is idle, nothing is lost. The system breathes like a living organism - inhaling sunlight, exhaling power, and maintaining balance.

The Canal as the Circulatory Path

The **Pan-African Canal** functions as the bloodstream of this system. It carries not only water but energy in multiple forms:

- Pipelines for hydrogen and ammonia run along its banks.
- Ships powered by the same fuels move goods and people.
- Floating solar stations feed nearby grids and irrigation systems.
- Ports become hubs for energy exchange and storage.

Just as rivers sustain ecosystems, the canal sustains industries, agriculture, and cities. It unites inland energy centers with coastal trade routes, keeping Africa's renewable flow uninterrupted.

Integration Across Sectors

The circular energy system also bridges sectors that were once separated.

- **Agriculture** uses green ammonia fertilizer and powers its pumps with renewable electricity.
- **Industry** uses clean hydrogen to manufacture goods, sending its waste heat to homes.
- **Cities** recycle organic waste into energy that powers transport and public services.
- **Forestry** absorbs carbon while producing biomass for energy recovery.

Every sector contributes to the circle and benefits from it. The waste of one becomes the wealth of another.

Storage and Stability

Circular systems need memory - the ability to hold energy until it is needed. The **Afar salt caverns** serve as natural memory banks, storing hydrogen as potential energy. Batteries, microgrids, and smart-grid software act as short-term memory, balancing fluctuations between regions. Together, they make the system flexible and stable.

Technology and Intelligence

Technology alone is not enough; it must be guided by intelligence. Digital systems monitor flows of electricity, hydrogen, and water across borders, ensuring equilibrium and efficiency. Artificial intelligence can predict demand patterns, optimize resource use, and minimize waste.

But true intelligence comes from philosophy - the belief that progress should serve life, not consume it. In Africa's circular system, technology follows wisdom, not the other way around.

People as Part of the Circle

The circular energy model only works when people participate. Farmers, technicians, students, and local cooperatives all become stewards of energy. Each community can produce, manage, and recycle its own resources, staying connected to the continental network. Ownership becomes shared, and sustainability becomes cultural.

A Living Energy Civilization

The circular system is more than infrastructure - it is a way of life. It teaches patience, efficiency, and balance. It transforms Africa into a continent that powers itself without harming itself, a civilization that grows by renewing rather than consuming.

Through **GERD+**, the **Pan-African Canal**, and the circular economy that surrounds them, Africa can become the world's first truly renewable civilization - where energy behaves like nature: **endless, balanced, and alive.**

8.3 Integrating Recycling with Power Generation

Recycling is more than a process of recovery - it is an act of intelligence. It shows that we understand the value of what already exists. When Africa integrates **recycling with power generation**, it completes the circle between human creation and nature's regeneration. Nothing leaves the system unused; every material, every molecule, and every idea finds a second life.

The **GERD+** framework and the **Pan-African Canal** offer a continental foundation for this integration. With renewable power in abundance, recycling can move from small-scale activity to a continental industry - one that transforms waste, saves energy, and builds jobs for millions of people.

Energy as the Engine of Recycling

Recycling needs power - for collection, processing, sorting, and remanufacturing. In many regions, that power is expensive or polluting, making recycling difficult to sustain. GERD+ changes that equation by supplying **clean, affordable energy** for recycling industries across Africa.

Solar electricity and hydrogen from GERD+ can run waste-sorting plants, metal and glass furnaces, and plastic recovery centers without emitting carbon. Energy derived from water and light breathes new life into what society once threw away.

With this clean energy base, recycling becomes a force for creation, not merely correction. It moves from charity to industry, from cleanup to innovation.

From Waste Streams to Resource Streams

Every African city produces waste - plastic bottles, metals, glass, textiles, paper, and electronics. These are not burdens; they are **urban mines**, rich in recoverable material. By connecting recycling systems to renewable grids, Africa can transform these waste streams into resource streams.

Hydrogen-powered collection vehicles can transport recyclable materials to local processing hubs. Biogas from organic waste can supply energy to recycling plants. Plastics can be reprocessed into construction materials. Metals can be melted and recast using clean electricity from GERD+.

Each recycled product reduces the demand for new extraction, protecting Africa's land and ecosystems. The continent gains not only materials but also time - the time that nature needs to renew itself.

Industrial Symbiosis

Recycling becomes most powerful when industries collaborate. One factory's waste heat can power another's process. Scraps from metal manufacturing can feed into construction materials. Excess steam from hydrogen production can dry agricultural goods.

This approach - known as **industrial symbiosis** - fits naturally within the GERD+ energy network. Shared energy corridors and canal-based industrial zones make it possible for factories to exchange resources efficiently. In this ecosystem, nothing stands alone. Every industry becomes part of a living industrial forest - diverse, self-sustaining, and interdependent.

Circular Construction

Recycling also reshapes how Africa builds. Construction and demolition waste can be converted into new materials - cement substitutes, bricks, and panels. Using energy from GERD+, these processes emit a fraction of the carbon of traditional manufacturing. The result is a **clean construction economy** that builds cities without burdening the earth.

Skyscrapers, housing complexes, and canal infrastructure can all use materials made from recycled products. The same energy that powers Africa's light can build its homes.

Technology and Digital Tracking

To integrate recycling effectively, Africa needs intelligent systems for tracking materials. Digital platforms can monitor the flow of resources from production to use and back to recycling. Smart labels and blockchain-based certification can verify recycled content and ensure transparency.

Powered by renewable energy, these technologies create trust in the circular economy - proving that sustainability is not just claimed but measured.

Community Participation

No recycling system succeeds without people. The circular model empowers communities to participate directly. Waste pickers and informal recyclers, long ignored, become trained professionals operating within organized cooperatives. Their work is recognized, formalized, and respected.

With reliable energy, small recycling centers can thrive in villages and towns. Youth cooperatives can manage local collection networks, powered by solar electricity or biogas. Communities no longer see waste as a problem to remove, but as a resource to manage.

Energy and the Philosophy of Renewal

When recycling and power generation merge, energy itself becomes a teacher. It shows that every ending is a beginning, that creation never truly stops. The lights powered by recycled materials remind us that progress does not require depletion - only imagination.

Africa's recycling-powered energy system is not about copying global models; it is about living wisely within its own rhythm. It turns cities into gardens, industries into ecosystems, and waste into wonder.

When every object finds another use and every factory shares its energy, Africa will have achieved what the world still seeks - **a civilization that grows by renewing, not by consuming.**

8.4 African Cities as Energy Producers

For much of modern history, cities have been seen as consumers - places that take in food, fuel, and materials, and release waste, heat, and pollution. But the cities of Africa's future must break that pattern. They must become **producers** - of clean energy, clean air, and clean knowledge. The **GERD+** system and the **Circular Africa** framework together make this shift possible, turning cities from energy sinks into self-sustaining power ecosystems.

The City as a Living Organism

A city is a body. Its roads are arteries, its people are cells, its industries are organs. When it runs only on imported fuel, it lives in imbalance. But when it produces its own energy from sunlight, waste, and water, it becomes self-healing - a living organism that regenerates what it consumes.

Africa's cities, both large and small, have everything they need for this transformation: abundant sunlight, organic waste, human creativity, and growing infrastructure. The key is to connect these elements intelligently, powered by renewable systems from GERD+ and local circular networks.

Waste-to-Energy and Recycling Power Plants

Urban waste is no longer a burden when converted into energy. Biogas plants near markets can transform organic waste into electricity and cooking gas. Municipal waste treatment centers powered by renewable energy can produce biohydrogen.

Each district becomes a microgrid, balancing energy generation and consumption. Instead of smoke and smell, cities will emit light and life.

Hydrogen Mobility in the City

Hydrogen-powered buses, taxis, and trucks can move people and goods quietly and cleanly. Refuelling stations supplied by hydrogen pipelines from GERD+ or local electrolysis units along the **Pan-African Canal** will make zero-emission transport practical and affordable.

This transformation of mobility changes not only air quality but also perception - the city becomes a symbol of progress, a place where innovation and health coexist.

Urban Agriculture and Circular Water Systems

Energy production in cities can also nourish food production. Renewable-powered irrigation systems and vertical farms can grow vegetables using recycled water and waste heat. The byproducts of these systems - organic residues and wastewater - return to energy and fertilizer cycles.

This creates a **self-feeding circle**: water supports food, food supports energy, and energy supports life. The same flow that powers the dam and canal can sustain the heart of the city.

Employment and Innovation

Transforming cities into energy producers will create millions of jobs - technicians, engineers, urban farmers, and digital system operators. Youth can manage solar installations, hydrogen logistics, recycling centers, and electric vehicle networks.

Each city will have its own **Energy and Circularity Hub**, where research, entrepreneurship, and training come together. These hubs will help cities share knowledge, learn from each other, and design locally adapted solutions.

Cultural and Psychological Renewal

When a city begins to produce its own power, its people begin to think differently. Pride replaces dependency. Clean air replaces noise and pollution. Public spaces once neglected become centers of learning and community life.

Children grow up seeing sunlight not as heat to escape but as a resource to embrace. Neighbourhoods celebrate the turning on of local grids the same way past generations celebrated rainfall - as a gift of life.

Cities as Lights of the Continent

If every African city generates its own clean energy, the continent will become a constellation of light - not just visible from space, but felt in the daily wellbeing of its people.

Each city, from Addis Ababa to Lagos, from Nairobi to Dakar, can become both consumer and contributor, both home and powerhouse. The energy that flows from GERD+ and the Pan-African Canal will not stop at borders or power stations - it will flow into every street, every building, and every dream.

Africa's cities will no longer be seen as places that drain resources. They will be seen as **living systems that give back** - radiant, productive, and alive.

When a continent's cities start producing more than they consume, its people begin believing more than they doubt. That is when transformation becomes unstoppable.

8.5 The Future of Urban Sustainability

The cities of tomorrow will be the mirrors of our wisdom. If they are polluted, it means we have lived without care. If they are green, peaceful, and bright, it means we have learned to live in harmony with what sustains us. Africa's future cities must embody this wisdom - places where technology and nature grow side by side, powered by clean energy and guided by the principle that **progress should never destroy life**.

The **GERD+** and **Pan-African Canal** systems make this future possible. Together, they provide the clean power, the water, and the philosophy needed to redesign Africa's urban landscape into a network of **living cities** - cities that breathe, grow, and renew themselves.

Designing with Nature, Not Against It

Urban sustainability begins with design. Cities must follow the rhythm of the land - using natural airflow, daylight, and green space to reduce energy use. Water from the canal and local reservoirs can be recycled through wetlands and purification systems that double as parks.

Buildings can be constructed using eco-materials, powered by solar and hydrogen systems, and cooled by smart design rather than by wasteful machines. Streets shaded by trees reduce heat and clean the air. The city becomes a **forest of innovation**, where every structure contributes to life.

Energy Independence for Every District

Each city district will operate as a **micro-energy ecosystem**, linked to the continental grid but able to sustain itself independently. Floating solar systems on lakes, rooftop panels, and waste-to-energy plants will power neighbourhoods, schools, and hospitals.

Hydrogen and ammonia storage will provide stability during cloudy seasons, while smart grids balance supply and demand automatically. In this way, cities become resilient - never dependent on one source or one region.

Energy independence also creates security. Power no longer becomes a political weapon or an economic burden. It becomes a shared right - quietly serving everyone, every day.

Clean Mobility and Silent Streets

The future African city will move without noise or smoke. Hydrogen-powered buses, trams, and cargo vehicles will glide through clean streets. Pedestrian walkways, cycling routes, and green corridors will replace congested roads.

Hydrogen refuelling stations powered by GERD+ and the canal network will make zero-emission transport possible even for heavy-duty industries. The hum of engines will give way to the soft rhythm of electric motion. The air will once again smell of rain, not of fuel.

Circular Water and Waste Systems

Urban sustainability depends on how we manage what we discard. Cities of the future will close the loop on waste and water. Every drop used will be treated and reused. Every kilogram of waste will be recycled, composted, or converted into energy.

Neighbourhood recycling centers, biogas facilities, and greywater treatment plants will become standard components of city planning. Powered by renewable electricity, they will transform waste into employment and pollution into productivity.

Urban Food and Green Space

Sustainable cities must also feed themselves. Rooftop gardens, vertical farms, and canal-side greenhouses will supply fresh produce powered by renewable energy. Organic waste will return as fertilizer, and clean irrigation systems will circulate through the city like veins of life.

This urban agriculture will reconnect people with the soil, making cities not the opposite of nature, but extensions of it. Children will grow up knowing how food grows, not just where it is sold.

Social Equity Through Design

A sustainable city must be inclusive. The energy transition must not serve only the privileged; it must uplift all. Affordable housing powered by solar microgrids, accessible transport, and public green spaces will ensure that every citizen participates in the new urban order.

The **GERD+** model makes this inclusion possible by lowering energy costs and expanding access to reliable power. Equality becomes not just a moral ideal, but a practical outcome of smart systems.

The African City as a Teacher

When built wisely, the African city of the future will not copy others - it will **teach**. The world will look to Africa for models of cities that thrive in balance with water, sunlight, and community.

A city powered by hydrogen, cooled by trees, and fed by recycled water is not science fiction. It is a natural evolution of Africa's own heritage - a modern reflection of the traditional villages that lived in rhythm with the seasons.

From Survival to Serenity

The future of urban sustainability is not about endless expansion; it is about peaceful coexistence. It is a future where growth is measured not by size but by harmony. Where clean air, quiet streets, and creative citizens define progress more than skyscrapers and traffic.

When Africa's cities reach this stage, the continent will no longer chase modernity - it will define it.

Each city will stand as a symbol of what Africa believes: that development can heal, that energy can purify, and that life can thrive without harm.

The cities of the future will not just house people; they will inspire them. They will whisper the same message as GERD+ and the Pan-African Canal: **to build is to care, to grow is to renew, and to live well is to live in balance.**

9.0 FORESTS, CARBON, AND THE CLIMATE EQUATION

A living continent breathes. Its lungs are its forests, its blood the rivers, and its breath the wind that moves across its hills. If energy gives life to Africa's cities and industries, **forests give balance to its soul**. The future of Africa's development cannot stand on technology alone - it must rest on the harmony between power and nature.

The **GERD+** and **Pan-African Canal** systems are built upon this balance. They provide the clean energy that makes deforestation unnecessary, and the water that allows trees to grow again where deserts once spread. They are not just tools for progress; they are instruments of restoration.

Africa's forests are vital to the planet's health. The Congo Basin alone absorbs billions of tons of carbon every year. From the highlands of Ethiopia to the savannas of Kenya, from the Sahel to the rainforests of Central Africa, these ecosystems regulate rainfall, anchor soil, and moderate climate. To protect them is to protect ourselves.

But protection alone is not enough. The future demands **regeneration** - planting new forests, restoring degraded lands, and building a circular relationship between trees, energy, and human life.

Forests as Climate Infrastructure

Forests are more than scenery. They are living power stations that operate silently, day and night. Each leaf captures carbon, each root stores water. In a warming world, trees are not a luxury - they are **infrastructure**.

The **GERD+** initiative recognizes forests as part of Africa's energy system. Clean hydropower and hydrogen production reduce the need for wood-based fuel, protecting forests from being cut for charcoal or firewood. Floating solar fields, waste-to-energy plants, and efficient stoves ensure that rural communities no longer depend on deforestation for survival.

As energy becomes renewable, forests can become renewable too.

The Carob and the Moringa: Trees of Renewal

Africa's reforestation movement can draw strength from its own native species - hardy, intelligent trees that thrive in difficult climates. The **Carob tree**, symbolic of patience and abundance, can grow in arid lands, restoring soil and providing shade, fruit, and livestock feed. The **Moringa tree**, known as the "tree of life," offers nutrition, medicine, and oil. Together, they form the backbone of **living landscapes** - forests that feed both nature and people.

Planting these trees along the **Pan-African Canal** will create continuous green corridors - belts of life that absorb carbon, stabilize microclimates, and beautify the land. Every kilometer of the canal can become both an irrigation route and a reforestation line.

Carbon Balance Through Regeneration

In the global climate equation, Africa holds a unique role. The continent emits little but absorbs much. By expanding its renewable energy systems and reforesting its lands, Africa can become the world's largest carbon sink - a **continent that cools the planet** while powering itself sustainably.

A coordinated program linking GERD+, the canal, and forestry initiatives could generate vast "green carbon credits," but more importantly, it would restore ecological dignity. Carbon would not just be counted; it would be **cultivated** - through living forests rather than market instruments.

Hydrogen and Carbon Neutrality

The clean hydrogen produced by GERD+ ensures that Africa's energy systems emit almost no carbon. Hydrogen-based industries - fertilizers, shipping, and manufacturing - replace fossil fuel processes, bringing the continent closer to carbon neutrality.

When this clean energy is paired with reforestation, the result is **carbon positivity**: Africa would absorb more carbon than it releases, reversing global warming trends while sustaining growth.

Forests as Economic Systems

A thriving forest economy does not destroy trees - it grows them. Communities along reforestation corridors can earn livelihoods through seed collection, nursery management, and sustainable harvesting of fruits, oils, and resins. Green industries powered by renewable energy can process these forest products locally, adding value without exploitation.

Eco-tourism, research, and education will flourish around these green landscapes, reconnecting people to the land that sustains them.

The Spiritual Dimension

Forests also renew the spirit. They remind us of patience - the slow strength that comes from growing quietly but deeply. African wisdom has always revered trees as symbols of connection between earth and sky. To plant a tree is to believe in tomorrow; to protect a forest is to honour the ancestors who lived in balance with nature.

In this sense, every solar field, every hydrogen plant, and every dam must also be seen as a **forest of ideas** - designed not to dominate, but to coexist.

A Continent That Breathes Again

Africa's transformation will not be measured only in electricity or income, but in **fresh air, clean water, and living soil**. When forests return, rainfall will balance, rivers will refill, and life will stabilize.

The GERD+ and Pan-African Canal projects are not separate from this goal - they are its engines. Energy gives the tools; forests give the equilibrium.

The future of Africa's climate leadership is not written in emissions charts; it is written in leaves. Every tree planted under the light powered by GERD+ will stand as proof that Africa's progress can heal the planet it inhabits.

A continent that powers itself cleanly and breathes freely will no longer be called developing - it will be called **wise**.

9.1 Restoring the Green Cover of Africa

To restore Africa's green cover is to restore its balance. The forests, grasslands, and wetlands are not separate from our progress - they are the foundation of it. When the land is green, the rivers flow, the rains return, and the soil remembers how to feed. When the land is bare, everything suffers. Restoring the green cover of Africa is not just an environmental duty; it is an act of survival, peace, and wisdom.

The **GERD+** and **Pan-African Canal** systems make this restoration not only possible but practical. They provide what nature needs most to heal - **water, energy, and time.** With irrigation powered by renewable energy and guided by intelligent design, the deserts of despair can once again bloom.

From Degradation to Regeneration

Over the decades, large areas of Africa have been degraded by overgrazing, deforestation, and erratic climate. The Sahel has moved southward, rivers have shrunk, and once fertile soil has turned to dust. Yet the same land that suffered can recover if given the chance. The seeds of renewal already lie within it.

With the water management capacity of the **Pan-African Canal**, the soil can be rehydrated; with clean electricity from **GERD+**, the irrigation can be continuous; with the participation of local communities, the landscape can be reborn.

Where drought once ruled, orchards can grow. Where dust once moved, forests can stand.

Reforestation Corridors

The canal provides a natural backbone for **reforestation corridors** - continuous green belts stretching across countries and climates. Along its banks, trees like **Carob**, **Moringa**, and indigenous acacias will be planted systematically. These trees are chosen not only for their resilience but for their purpose: to restore soil, provide shade, and sustain life.

Each corridor becomes more than a line of trees - it becomes a **living infrastructure** of carbon absorption, wind protection, and economic opportunity. The roots stabilize the canal's banks, the leaves cool the air, and the shade creates microclimates where food crops and biodiversity thrive.

Energy for Regeneration

Reforestation is often limited by lack of resources: water pumps need fuel, nurseries need light, and farmers need tools. GERD+ changes this reality. With abundant renewable energy, irrigation systems can run continuously without emissions. Solar-powered nurseries can operate in remote areas, producing millions of seedlings each year.

Hydrogen-based machinery and electric vehicles can transport water and materials efficiently across the greenbelt projects. The same energy that powers cities can power forests.

Local Ownership and Participation

Forests thrive when people protect them. The reforestation effort will rely on **community ownership**, giving local families rights and incentives to manage the land sustainably. Each village along the canal can care for its section of the green corridor - planting, watering, and harvesting non-timber forest products such as fruits, seeds, and medicinal plants.

Education programs will train youth in agroforestry, soil restoration, and water conservation. Reforestation will no longer be an external project but a local pride. When people see themselves as guardians rather than beneficiaries, trees become family.

Water and Tree Synergy

The Pan-African Canal's irrigation network will distribute water to green corridors and nearby farms. Instead of short-term planting campaigns, this approach ensures long-term sustainability. Trees grow alongside crops, providing shade and retaining moisture. This integrated approach - **tree-crop coexistence** - increases yields, restores fertility, and protects biodiversity.

Trees and water are not separate systems; they are one. When trees grow, the air cools and moisture returns; when water flows, trees thrive. This cycle is the foundation of Africa's resilience.

Carbon and Climate Benefits

Reforesting degraded land can sequester billions of tons of carbon annually. Africa can become the world's leading region for natural carbon capture - not through offsets traded in distant markets, but through **living carbon systems** that benefit its people directly.

As forests expand, rainfall patterns stabilize, hydrological cycles strengthen, and the continent's water security improves. This in turn supports hydropower and agriculture, completing a circle of renewal - **forests feeding energy, energy feeding forests.**

A Symbol of Hope

Every seed planted is a statement of faith. Faith that tomorrow will be greener than today. Faith that Africa can heal itself with its own hands.

Restoring the green cover of Africa is not only about trees; it is about restoring confidence - the belief that the land we once called barren still remembers how to bloom.

With GERD+ powering irrigation and the Pan-African Canal carrying water across borders, Africa can turn its scars into gardens.

When forests rise again, rivers will sing again. And when the land is green, the people will breathe freely - for a continent that restores its trees restores its soul.

9.2 The Carob and the Moringa: Africa's Living Batteries

In every forest, there are trees that do more than grow - they give. They hold water, store sunlight, feed the soil, and nourish life long after they are gone. The **Carob** and the **Moringa** trees are two such gifts. They are not just plants; they are **living batteries** - natural systems that capture energy, store nutrients, and regenerate the land. For Africa, they symbolize how growth and wisdom can coexist.

The Carob: The Tree of Patience and Plenty

The **Carob tree**, ancient and resilient, has been called *the golden tree of dry lands*. It thrives where others fade. Its roots reach deep into the earth to find water, binding the soil and preventing erosion. Its thick canopy provides shade in harsh climates, and its pods - rich in sugar and protein - feed both people and livestock.

In the **Pan-African Canal green corridors**, the Carob becomes a guardian of stability. Its roots hold the canal's edges firm, its leaves reduce evaporation, and its fruits provide income for local communities. The Carob does not demand much - it gives quietly, over centuries. It is a tree that teaches patience: that real strength grows slowly, but lasts forever.

Energy and food flow from its presence. Processed with renewable electricity from **GERD+**, Carob pods can become natural sweeteners, bio-based fuels, or ingredients for animal feed. The tree becomes an example of how agriculture and industry can unite in circular harmony - producing nourishment without destroying nature.

The Moringa: The Tree of Life

If the Carob represents patience, the **Moringa** represents abundance. Every part of the Moringa tree - leaves, seeds, flowers, and bark - serves life. Its leaves are a rich source of protein, iron, and vitamins. Its seeds produce oil that can purify water and fuel lamps. Its roots improve soil fertility. The Moringa grows fast, even in dry zones, restoring degraded land within seasons rather than years.

Planted along the **GERD+–Canal energy corridor**, Moringa trees become small solar factories of nature - capturing sunlight and converting it into food, medicine, and clean air. Their leaves feed families; their seeds filter drinking water; their branches anchor biodiversity.

Powered by Africa's renewable energy systems, Moringa processing centers can create nutritional powders, natural oils, and health products for both local and global markets. This is the essence of sustainable prosperity - using what grows naturally to power what humanity needs.

Trees as Living Energy Storage

Both the Carob and the Moringa demonstrate that nature has its own way of storing energy. A tree absorbs sunlight and carbon, turning them into biomass - energy preserved in living form. When grown across millions of hectares, these trees form **living batteries** that balance the planet's climate.

Unlike machines, they do not need replacement or maintenance; they only need protection. They store carbon instead of emitting it. They release oxygen instead of pollution. Their shade cools the ground; their roots guide the rain.

These trees, powered indirectly by the clean systems of **GERD+**, complete Africa's energy equation. While turbines generate electricity, trees generate balance. One powers the grid; the other powers the earth.

The Human Connection

The Carob and the Moringa are also teachers. They remind Africans of their ancestral relationship with the land - a partnership built on respect, not conquest. Planting them revives both ecology and culture. Communities involved in these green corridors will see trees not as resources to cut, but as companions to keep.

Women's cooperatives can process Carob pods and Moringa leaves into high-value products. Youth cooperatives can manage nurseries and plantations, earning both income and purpose. Schools can include tree cultivation in their curricula - teaching children that the future begins in the soil beneath their feet.

A Symbol for Circular Africa

In the vision of **Circular Africa**, the Carob and Moringa embody the continent's philosophy: that progress is circular, not linear; that every act of creation must also be an act of care. These trees store the sun's energy the same way GERD+ stores the river's - quietly, efficiently, and endlessly.

When planted across the canal's route, the highlands, and the drylands, these species will transform Africa's landscape into a tapestry of green resilience. They will power the atmosphere with oxygen, cool the air, and rebuild the soil.

A Future Rooted in Living Energy

The Carob and the Moringa are not simply part of reforestation; they are symbols of Africa's spiritual and ecological rebirth. They show that true progress does not replace nature - it grows with it.

As GERD+ lights the continent with electricity, these trees will light it with life. As turbines spin and canals flow, the forests will hum in silent cooperation. Together, they will form an Africa that no longer extracts to survive, but regenerates to live.

When a child plants a Carob or a Moringa, they plant a power station - one that runs on sunlight, compassion, and time. And when millions of such trees rise, the continent itself will breathe as one - steady, strong, and self-renewing.

9.3 Tree-Based Carbon Systems for Energy Offsets

The cleanest energy system in the world is not powered only by turbines or solar panels - it is powered by balance. For every watt of energy we produce, nature must be able to recover what we take. That recovery comes from trees. Forests are the Earth's oldest and most efficient **carbon systems**, and they are essential partners in Africa's energy renaissance.

While **GERD+** and the **Pan-African Canal** generate and move renewable power, the continent's forests ensure that energy growth remains climate-positive. They store carbon, protect soil, and regulate the rainfall that sustains hydropower. Together, they form a complete equation of energy and ecology - a cycle where production, protection, and regeneration are inseparable.

The Logic of Tree-Based Carbon Systems

In traditional economics, energy production adds carbon to the atmosphere. In Africa's new system, trees **subtract** it. By planting and protecting vast green corridors linked to the GERD+ and canal networks, the continent can offset any residual emissions from construction, industry, and transport.

Each tree acts as a **living carbon battery**, absorbing carbon dioxide and releasing oxygen. A mature forest of Carob, Moringa, acacia, and indigenous species can capture thousands of tons of carbon per hectare over its lifetime. When multiplied across the continent, the result is a vast natural storage system - silent, steady, and endlessly renewable.

Designing Energy–Carbon Symbiosis

Under the GERD+ framework, every major energy project can include a **carbon restoration zone** - a designated area where trees are planted, monitored, and managed to balance the emissions from construction and operation.

For example:

- A solar or hydrogen plant funds and maintains its own surrounding forest belt.
- The Pan-African Canal's irrigation networks double as tree nurseries.
- Floating solar arrays use shaded reservoirs that reduce evaporation and encourage nearby vegetation growth.

This integration ensures that carbon removal is not a side project but an **inherent part of design**. Energy and ecology become two faces of one system.

Digital Carbon Monitoring

Modern technology makes this vision measurable. Satellites, sensors, and artificial intelligence can track forest health, biomass density, and carbon absorption in real time. Africa can lead the world in transparent carbon accounting - not by selling offsets to others, but by building a **continental carbon register** that verifies Africa's contribution to the planet's cooling.

This data will empower African nations to set their own value for carbon - one that reflects the real cost of preservation, not prices decided elsewhere. It turns forests into economic assets that support communities while protecting the climate.

Hydrogen and Carbon Neutrality

The hydrogen produced by GERD+ already has near-zero emissions, but tree-based carbon systems make it **carbon-negative**. That means every kilogram of hydrogen exported carries not only energy but also a contribution to global restoration. African hydrogen becomes **green in**

both colour and conscience - a symbol of energy that heals rather than harms.

The same trees that offset emissions also secure the water cycles that hydropower depends on. Forests in Ethiopia's highlands, the Congo Basin, and the Sahel act as natural regulators for rainfall and river flow, ensuring the sustainability of GERD+ and other dams.

Community Carbon Cooperatives

Tree-based carbon systems also provide livelihoods. Communities can form **carbon cooperatives** that manage forest corridors, measure growth, and earn revenue from verified carbon storage. These cooperatives can use renewable energy from GERD+ to power irrigation, processing, and monitoring equipment, ensuring that climate protection also brings economic opportunity.

When farmers plant trees on their land, they are not losing space - they are gaining security. Trees improve soil fertility, protect crops from wind, and provide shade for livestock. Carbon revenue becomes an additional income stream, turning environmental care into a profession of pride.

Beyond Offsetting: A Philosophy of Balance

True sustainability does not mean compensating for harm; it means preventing it. Africa's tree-based carbon systems go beyond offsetting - they create **permanent balance** between humanity and nature. Forests are not planted as excuses but as extensions of energy systems that already work in harmony with the Earth.

This is the moral difference between a carbon market and a living carbon civilization. The former trades guilt; the latter grows grace.

The Continental Carbon Canopy

When all these green corridors and forest belts connect, they will form a **Continental Carbon Canopy** - a living network of lungs that covers Africa from the Sahel to the Cape, from the Nile to the Congo. Under this canopy, cities breathe cleaner, rivers run steadier, and the air carries renewal instead of smoke.

The world will one day recognize Africa not only as a source of clean energy, but as the **continent that restored the planet's balance**. The GERD+ turbines may hum with the rhythm of progress, but above them, the forests will whisper the wisdom of preservation.

Energy that gives life and forests that protect it - together, they form the only true currency of the future: **balance.**

9.4 Forestry, Hydrogen, and Circular Growth

The true beauty of Africa's transformation lies in connection - the ability to make every part of life feed another. In the old industrial world, energy, nature, and economy were separate. One extracted, another suffered, and the third grew distorted. Africa's future must be different. It must be **circular**, where every source of growth gives back more than it takes. The relationship between **forestry and hydrogen** shows how this can be done - how energy can nurture nature, and nature can strengthen energy.

Hydrogen Sustains Forests

Traditional energy has been one of the greatest causes of deforestation in Africa. Millions of households still rely on wood and charcoal for cooking and heating. This daily necessity consumes vast areas of forest each year. By providing clean, affordable hydrogen energy - and electricity from **GERD+** - Africa can finally release its forests from this burden.

Hydrogen can replace firewood and charcoal in homes and small industries. It can fuel clean stoves, boilers, and local generators. This single shift could save millions of trees annually. When energy becomes clean, forests are free to grow again.

Forests Support Hydrogen

Forests, in turn, play a vital role in sustaining the hydrogen economy. Trees regulate rainfall, stabilize soils, and protect the watersheds that feed hydropower. Without healthy forests, rivers dry, dams lose capacity, and energy stability fades. Planting and protecting trees is therefore not an environmental side task - it is **energy security**.

The forest canopy keeps the air cool, reducing evaporation from reservoirs such as GERD+. Its roots prevent erosion that can silt up dams. Its presence ensures that the water cycle - and therefore the energy cycle - continues.

Biomass and the Hydrogen Chain

In a circular system, even forest by-products become part of the energy loop. Fallen branches, pruned wood, and agricultural residues can be collected to produce **biohydrogen** and **biogas** through gasification or fermentation. These clean fuels complement the renewable hydrogen generated from hydropower and solar energy.

Nothing is wasted:

- Forest residues become energy.
- Hydrogen by-products (oxygen and water) return to nurture the land.
- Waste heat from hydrogen plants powers nurseries and greenhouses.

It is a perfect circle - **trees feeding energy, energy feeding trees**.

Hydrogen for Reforestation Logistics

Hydrogen energy also enables the logistics of large-scale reforestation. Electric and hydrogen-fuelled vehicles can transport seedlings, pump irrigation water, and power agricultural machinery in remote areas without polluting. Portable hydrogen units can supply electricity for reforestation camps, monitoring stations, and nurseries.

The same clean energy that powers Africa's industries will power the quiet work of renewal.

Industrial Symbiosis: Wood and Wisdom

Forestry, powered by renewable energy, can become a foundation for sustainable industry - producing wood products, paper, furniture, and natural fibers while maintaining ecological balance. Modern sawmills and processing plants can operate on electricity or hydrogen, producing zero emissions.

Waste from these industries feeds bioenergy systems, creating jobs while keeping carbon in circulation rather than in the atmosphere. Every forest-related enterprise becomes an **eco-industry**, where waste is raw material and growth is regeneration.

Circular Growth in Practice

The partnership between hydrogen and forestry embodies Africa's circular future:

- **Hydrogen** replaces wood fuels, preserving forests.
- **Forests** protect water and climate, sustaining hydrogen production.

- **Biohydrogen** and **biogas** close the loop, turning organic matter into power.
- **Communities** earn livelihoods from both tree management and energy generation.

This model creates prosperity without depletion, industry without pollution, and progress without regret.

Forests as Living Economies

The circular growth approach turns forests from passive conservation zones into **living economies**. Managed responsibly, they become productive without being destructive. Communities along the **Pan-African Canal** and GERD+ corridors can participate in forestry cooperatives - cultivating, harvesting, and replanting in harmony with nature.

Trees become renewable assets, not one-time resources. Each harvest funds the next planting, creating continuous employment and wealth.

Hydrogen and the Spirit of Renewal

Hydrogen itself represents renewal - the smallest element, endlessly recyclable, burning clean, and returning to water. It reflects the same philosophy that governs a tree: transformation without harm. Together, hydrogen and forestry form the twin symbols of Africa's living civilization - one made of molecules, the other of leaves, both flowing in the same cycle of life.

Toward a Circular Continent

When Africa's forests are protected by clean hydrogen and its energy is balanced by trees, the continent achieves a form of harmony the world has forgotten - **growth that regenerates**. The rivers that power the dams, the sunlight that powers the panels, and the forests that breathe for the air all work as one system.

This is not just sustainability; it is sanctity - the return to a sacred order where nothing lives alone.

Africa's renaissance will not only be measured by its cities and industries, but by the health of its forests. The day a hydrogen flame lights a village while trees grow stronger in the hills, the circle of circular growth will be complete.

9.5 A Continental Climate Responsibility

Africa's climate responsibility is not a burden - it is a calling. The continent has contributed the least to global warming, yet it carries the greatest potential to heal the planet. It holds one fifth of the world's forest carbon sinks, the largest solar belt on Earth, and more untapped renewable water than any other region. The path of **GERD+**, the **Pan-African Canal**, and **Circular Africa** is not simply about development - it is about assuming the role of **planetary caretaker**.

From Victimhood to Leadership

For too long, Africa has been seen as a victim of climate change - a continent to be rescued, not one capable of rescuing itself. That perception must end. The new Africa, rising through clean energy and restoration, does not ask for sympathy; it offers **solutions**.

Through the GERD+ framework, Africa proves that it can power itself and still cool the planet. Through reforestation and carbon restoration, it shows that prosperity and preservation are not opposites. Through the Pan-African Canal, it demonstrates that water can be shared without conflict.

Africa's climate responsibility is leadership - not reactive, but visionary.

Climate Justice Through Action

Real climate justice is not about blame; it is about balance. The nations most affected by climate change should also be the ones who show how to reverse it. By choosing renewable power over fossil fuels, and regeneration over extraction, Africa claims the moral and practical leadership the world needs.

GERD+ and related projects form the backbone of this justice. They prove that clean energy can drive industrialization without environmental debt. Africa will not repeat the mistakes of others; it will **rewrite the model**.

The Climate Dividend of Africa

When Africa restores its forests, builds hydrogen economies, and manages its energy wisely, the whole world benefits. The air becomes cleaner, global temperatures stabilize, and biodiversity expands. This is Africa's **climate dividend** to humanity - a gift created not through wealth, but through wisdom.

And yet, Africa's leadership in climate repair must also bring fair recognition. The continent's carbon absorption and renewable innovations must be valued globally. Climate finance should no longer be charity; it should be **partnership** - an exchange of equals contributing to the planet's survival.

Shared Responsibility, Shared Gain

Africa's climate systems are interconnected. The rain that falls in Ethiopia may feed rivers in Sudan; the forests of the Congo may influence weather in West Africa. Therefore, climate responsibility must be **continental**, not national.

Shared monitoring, reforestation corridors, and energy grids can help every country participate. The African Union can guide this effort, ensuring that no region carries the weight alone. This is not about dividing the carbon burden - it is about **uniting the carbon opportunity**.

Each nation becomes a guardian of one part of the ecosystem, knowing that its success sustains all others.

Technology and Tradition Together

Africa's climate leadership will emerge not from copying foreign systems but from combining **modern science with traditional wisdom**. Indigenous practices - terracing, rainwater harvesting, rotational farming - have long preserved balance in African landscapes. When joined with hydrogen technology, digital monitoring, and renewable energy, these ancient insights become even more powerful.

The result is a climate model rooted in African identity - technological yet spiritual, global yet grounded.

The Ethics of Restoration

Climate responsibility is also moral responsibility. It asks every generation to leave the Earth better than it received it. In Africa's philosophy, this duty already exists: land belongs to both ancestors and unborn children. To harm it is to break the chain of life; to restore it is to strengthen the bond.

By protecting its forests, rivers, and atmosphere, Africa lives according to its oldest values - respect, reciprocity, and renewal.

A Message to the World

When the world looks to Africa in the coming decades, it will not see a continent waiting for aid, but a continent lighting the way. It will see **GERD+** powering cities cleanly, forests expanding across drylands, and hydrogen ships leaving African ports with zero emissions.

Africa will show that the fight against climate change is not about fear, but about faith - faith in life's ability to renew itself when guided by wisdom.

This is Africa's message: that healing the planet begins with **living in balance**.

From Responsibility to Renewal

Climate responsibility is not a punishment; it is a privilege. It allows Africa to define the moral compass of the modern world. The continent that was once exploited for resources will now export renewal - through energy, food, water, and hope.

The GERD+ system and the Pan-African Canal together form Africa's climate covenant - a promise written in light and water. They declare that the continent will not only survive the climate era but **lead it**.

When the winds carry clean air from African forests, when rivers flow freely between green banks, when the lights of renewable cities shine from the Sahara to the Cape - the world will know that Africa has kept its promise.

It will have fulfilled not just its climate responsibility, but its destiny: to become the continent that gives back more life than it takes.

10.0 THE FUTURE AFRICA

The Africa that is coming is not an extension of today - it is something entirely new. It is an Africa that remembers its roots while designing its own destiny. An Africa that grows not by consuming what others create, but by creating what the world will come to need. This is **The Future Africa** - a living, connected, self-sustaining civilization powered by knowledge, kindness, and light.

The systems we have envisioned - **GERD+**, the **Pan-African Canal**, the **Circular Economy**, the **Forestry Corridors**, and the **Hydrogen Networks** - are not separate projects. They are the organs of one continental body. Together, they form an ecosystem of progress where energy, food, and water flow as one.

This future is not far. It is already beginning wherever Africans build with their own ideas, teach with their own vision, and design with their own rhythm.

A Future Built, Not Borrowed

For generations, Africa was asked to follow other paths - imported models of growth that ignored the soul of the continent. The Future Africa will not imitate; it will **originate**. Its progress will be designed from its rivers, its cultures, and its people.

GERD+ has already shown that Africa can create world-class systems without external dependence. The same will be true across fields - education, health, technology, and governance. Every nation will contribute its own light to a shared horizon.

The Future Africa will be built on **sovereign creativity** - innovation that begins in African classrooms, laboratories, and communities. The continent's success will no longer depend on what it receives, but on what it reveals.

Technology with a Soul

In the coming decades, Africa's technology will not be cold or mechanical. It will be guided by empathy, designed to serve life rather than replace it. Renewable energy, smart agriculture, and clean industries will reflect the rhythm of the continent's ecosystems.

Artificial intelligence, robotics, and biotechnology will not erase humanity; they will extend it - allowing people to work more wisely, not harder; to create without destroying. The continent's advancement will be measured not only in machines, but in **meaning**.

The technology of the Future Africa will listen to the land, respect the seasons, and speak the languages of its people.

A Continental Mindset

The borders that divide Africa today will fade in importance. Trade, education, and energy networks will connect people more closely than lines on a map. The African Union's vision of unity will take form not just in declarations but in daily life - in grids that cross nations, universities that share curricula, and transport systems that move freely from coast to coast.

The Future Africa will think as one - diverse but aligned, independent but interlinked.

An Economy of Regeneration

In the Future Africa, growth will not mean depletion. Every new factory, city, or farm will contribute to regeneration rather than extraction. Hydrogen will fuel industries; circular systems will recycle materials; forests will grow as fast as goods are made.

Economic power will be measured in **renewability** - in how much life a system can sustain, not how much it can consume.

Jobs will come from restoration, not pollution. Wealth will come from balance, not exploitation. Prosperity will be peaceful because it will be shared.

Education as the Foundation

The Future Africa will be educated not only in science, but in consciousness. Its schools will teach systems thinking, ecological ethics, and the wisdom of coexistence.

Children will grow up learning that water, soil, and energy are part of one living network - that every action must heal what it touches. Education will be less about memorizing facts and more about **understanding relationships**.

In every African classroom, the future will already be visible - in the curiosity of young minds connected to clean light and to purpose.

Governance for the Living System

Politics in the Future Africa will evolve from managing scarcity to managing abundance. Leadership will no longer revolve around control of resources but around stewardship of systems.

Governments will act as coordinators of flow - ensuring that energy, water, and opportunity move freely and fairly. Decision-making will be guided by knowledge and compassion, not power or fear.

This new governance model - inspired by the balance of ecosystems - will become the hallmark of Africa's maturity.

Cultural Confidence

The Future Africa will express its confidence through culture. Music, design, and storytelling will carry messages of unity, dignity, and innovation. The arts will not escape from reality; they will **shape** it - inspiring industries, influencing technology, and teaching values.

African languages will flourish in literature and science alike. Fashion and architecture will blend modern functionality with ancestral identity. The continent's creativity will not only inspire Africans; it will re-educate the world about beauty, simplicity, and balance.

Africa and the World

Africa's rise will not threaten the world - it will **stabilize** it. The continent's renewable power and carbon balance will cool the planet. Its food systems will feed millions. Its clean hydrogen will move global trade. Its moral clarity will remind humanity what progress truly means.

The Future Africa will not compete for dominance; it will lead through example - as the **continent that grows without greed, builds without destruction, and leads without arrogance**.

The Human Vision

At the center of it all will be the African human - confident, creative, and compassionate. The Future Africa will measure success not in skyscrapers but in smiles; not in profit margins but in purpose; not in consumption but in contribution.

Every family with food, every child with light, every community with peace - these will be the true indicators of advancement.

The Circle Complete

When the systems of GERD+, the Pan-African Canal, Circular Africa, and the continental forests all function together, the circle will be complete. Energy will flow like breath. Cities will grow like gardens. People will live as part of nature, not apart from it.

The continent will not just survive - it will **thrive**. It will become the model of what civilization can be when guided by wisdom instead of speed, cooperation instead of competition, and care instead of conquest.

Africa's future will not be borrowed from others; it will be **born from within** - patient, confident, and alive.

The Future Africa is not a dream. It is already awakening.

10.1 The Africa of Systems and Wisdom

The strength of a civilization is not measured by how many things it builds, but by how wisely it connects them. **The Future Africa** will not be defined by isolated achievements - it will be defined by systems that think, breathe, and grow together. This is the **Africa of Systems and Wisdom**: an intelligent continent guided not by haste, but by harmony.

From Projects to Systems

Africa's past development often revolved around projects - individual efforts that ended where they began. The new era replaces projects with **systems**. GERD+, the Pan-African Canal, the Circular Economy, and Continental Reforestation are not separate plans. They are pieces of one living network.

Energy flows into agriculture. Water sustains forests. Waste becomes value. Knowledge moves across borders. This interconnected design turns Africa from a continent of scattered potential into a **coordinated organism** of progress.

When Africa builds as a system, every success strengthens another. A canal that irrigates also transports. A dam that generates power also feeds hydrogen production. A tree that shades a village also balances the climate. Each part adds value to all others.

This is how the Africa of Systems begins - not through competition, but through **coherence**.

The Intelligence of Nature

The greatest teacher of systems is nature itself. In forests, nothing is wasted; every leaf, root, and drop of rain serves a purpose. In rivers, flow creates life through movement, not ownership. Africa's new civilization will follow this same logic.

GERD+ mirrors the natural rhythm of water; the Pan-African Canal mirrors the veins of the earth; Circular Africa mirrors the cycle of life itself. Technology will not dominate nature - it will **extend** its intelligence.

This is not a mechanical system but a **living one**, where human design learns from ecological wisdom.

Wisdom Over Speed

The world often mistakes speed for progress. The Africa of Systems will reject that illusion. It will grow patiently, layer by layer, with care for both outcome and origin. The true measure of wisdom is not how fast something can be built, but how long it can last.

This continent will choose longevity over urgency, depth over display. Its systems will be designed to serve generations, not decades. This slower, wiser pace will become Africa's greatest advantage in a world racing toward exhaustion.

Systems Thinking as Education

The schools of the future will not teach only facts; they will teach **relationships** - how water connects to soil, how energy connects to health, how economy connects to morality.

Students will learn to think in circles rather than lines. They will design communities that regenerate themselves. They will see that knowledge is not about controlling nature but about cooperating with it.

From early education to universities, Africa's learning will reflect its living systems - practical, holistic, and wise.

Wisdom in Governance

A continent of systems needs leaders who think like systems - those who see patterns, not positions; who understand that decisions ripple through everything.

Governance will therefore evolve from ruling to **guiding flow**. Policies will be designed to balance, not to command. Ministries will no longer work in isolation; energy, water, agriculture, and education will plan together.

This is governance by **integration**, not fragmentation - the wisdom of orchestration rather than domination.

The Spiritual Dimension of Systems

Wisdom is not only intellectual; it is moral. Systems fail when they lose their center - their purpose. The Africa of Systems will succeed because it will remember that all design must serve life.

African spirituality, rooted in connection to land and community, already understands this. Every river has a soul, every tree a spirit, every person a role in the harmony of existence. This ancient consciousness will become the ethical foundation of modern systems.

GERD+, the Canal, and the forests will thus stand not as monuments to human power, but as symbols of **spiritual balance** - technology guided by humility.

Collective Intelligence

In the Africa of Systems and Wisdom, intelligence will no longer be centralized in institutions alone. It will be **collective** - distributed across communities, regions, and generations. Farmers will share local insights with scientists. Engineers will learn from elders. Policy will listen to experience.

This cooperation between modern science and traditional understanding will create a uniquely African form of intelligence - grounded, practical, and deeply moral.

Sustainability as Culture

When systems thinking becomes part of culture, sustainability is no longer an effort; it becomes instinct. People conserve not because they are told to, but because it feels natural. They see the connection between action and consequence, between taking and giving.

In this future, sustainability will not be a policy but a **way of being** - expressed in every meal, every building, every invention.

Africa as the Teacher of Systems

The world that once taught Africa will soon come to learn from it. As other continents struggle with disconnection - between people and planet, between economy and ethics - Africa will offer a new paradigm: the civilization that works like nature.

It will teach that growth can coexist with grace, that power can be gentle, that progress can be circular. The Africa of Systems and Wisdom will stand as a living university for humanity.

A Continent Aligned with Itself

When systems align with wisdom, peace becomes natural. Conflicts fade because cooperation becomes logical. Poverty declines because value circulates. Despair disappears because purpose returns.

The Africa of Systems and Wisdom will be an Africa **in alignment with itself** - where the mind of technology, the heart of culture, and the soul of nature beat in the same rhythm.

This is the Africa that GERD+ was designed to inspire - not a continent of survival, but a continent of **synchrony**, where everything lives, learns, and grows together.

10.2 The Awakening of African Intelligence

The greatest revolution of the coming century will not be industrial or digital - it will be **intellectual**. It will not begin in laboratories or boardrooms, but in the quiet realization of a continent remembering its own mind. The **Awakening of African Intelligence** is not the rise of machines or algorithms; it is the rise of consciousness - the recognition that true intelligence is already rooted in the land, in the people, and in the patterns of nature itself.

Africa's intelligence has always been living. It is not stored in data servers but in forests, languages, rivers, and relationships. It flows like water and grows like trees - adapting, resilient, circular. What GERD+, the Pan-African Canal, and the Circular Africa system represent is not only technological advancement but the reactivation of this **living intelligence** - a harmony of thought that unites reason and rhythm, logic and life.

Remembering, Not Inventing

For too long, the world has treated intelligence as something Africa must acquire. But Africa does not need to be taught how to think; it needs only to **remember** how it already thinks. From architecture that mirrors ecosystems to governance systems based on consensus, African civilizations have long practiced complex, adaptive intelligence.

This awakening is not about borrowing foreign frameworks; it is about **reclaiming indigenous wisdom** and merging it with modern tools. The result will not be imitation, but evolution - an intelligence shaped by history, struggle, and faith.

Natural Intelligence

Nature itself is the original form of intelligence. Every tree, every cloud, every river expresses a kind of knowing - responding, adjusting, and creating balance. Africa's future systems, inspired by nature, will not operate on rigid rules but on **adaptive learning**, just as the land does.

GERD+ functions according to this same principle. It balances energy flow with river ecology, adapts to climate, and supports multiple purposes simultaneously - irrigation, power, and peace. This is not mechanical control; it is **ecological intelligence** made visible.

In the same way, the Pan-African Canal will act as a neural pathway, connecting the continent's economies and ecosystems - transmitting energy, water, and information across regions. Africa's infrastructure will behave like a living brain.

Collective and Cooperative Intelligence

In the new Africa, intelligence will not be concentrated in individuals or machines. It will be **shared**, just as knowledge once circulated through oral traditions. Communities will use both digital technology and local experience to manage energy, agriculture, and trade cooperatively.

Each village, each city, each research center will act as a node in a continental network of learning - communicating through open platforms powered by clean energy. The sum of this cooperation will form a **continental intelligence** greater than any single nation's capacity.

Artificial intelligence will assist, but human understanding will lead. Machines will analyze data; people will interpret meaning. Technology will amplify wisdom, not replace it.

The New African Mindset

The awakening of African intelligence also means freeing the mind from dependency. For generations, Africa was told to receive instructions rather than to write them. The new mindset rejects that passivity. It understands that innovation does not need permission - it needs **belief**.

African intelligence will no longer define itself by catching up to others. It will move in its own rhythm - guided by observation, humility, and long-term purpose. It will see that thinking slowly, deeply, and collectively is a strength, not a weakness.

This mindset will restore confidence to African research, universities, and governance. It will produce thinkers who design for context - engineers who understand rivers, economists who understand rain, and philosophers who understand soil.

Spiritual Intelligence

Beneath Africa's scientific awakening is a deeper kind of awareness - **spiritual intelligence**. It is the understanding that knowledge without compassion is hollow, and progress without balance is blind. African traditions have always recognized that wisdom must serve life, not dominate it.

The continent's future systems will reflect this sacred balance. Every energy plant, canal, and forest project will carry a moral foundation: to give more than it takes, to heal more than it harms.

When science and spirit work together, knowledge becomes grace - intelligent action guided by empathy.

Youth as Carriers of Intelligence

Africa's youth will lead this awakening. They are the generation that will merge tradition and technology, coding and culture. They will turn renewable energy into entrepreneurship, water management into design, and agriculture into innovation.

With access to light, information, and opportunity, they will no longer see education as escape but as **creation**. Every young African who solves a local problem will add to the continent's collective intelligence - a brain of a billion neurons, all thinking for the same purpose: renewal.

The Language of Understanding

As intelligence awakens, Africa will also rediscover the power of its languages. Indigenous tongues carry ecological memory - names for plants, rains, and landscapes that hold centuries of science. Restoring these languages means restoring data systems encoded in culture.

When African languages return to classrooms and research, the continent will begin to think again in its **own metaphors** - the metaphors of rain, soil, light, and life.

From Intelligence to Wisdom

True intelligence is not cleverness; it is **care**. It understands that knowing more means being responsible for more. The awakening of African intelligence will therefore lead naturally to wisdom - the capacity to act rightly, not merely to act efficiently.

GERD+ and the systems that surround it embody this wisdom. They show that power can flow without harm, that water can serve all, and that light can lead without blinding.

The Future Mind

Africa's awakening will produce a new kind of mind - one that sees connection everywhere, one that learns from nature and teaches the world. It will not divide science from spirit, nor progress from peace. It will show that intelligence, when rooted in compassion, becomes **consciousness**.

This is the next phase of Africa's evolution - the **Mind of the Living Continent**. It will not conquer; it will compose. It will not rush; it will resonate.

And as this intelligence expands, Africa will no longer be seen as the continent of untapped potential. It will be recognized as the **continent of understanding** - a place where knowledge lives in balance with kindness, and where wisdom shines brighter than power.

10.3 Africa as a Living Civilization

Civilizations rise and fall not because of what they build, but because of how they live. The future of Africa will not be measured by its towers, its GDP, or its highways. It will be measured by the way it harmonizes energy, nature, and humanity - by how it **lives as a civilization**, not merely as an economy.

Africa as a Living Civilization is the natural conclusion of everything GERD+, the Pan-African Canal, and Circular Africa represent. It is a society that behaves like an ecosystem: circular, balanced, and self-renewing. It does not extract life to sustain progress - it uses progress to sustain life.

The Soul of Civilization

Every true civilization has a soul - an invisible rhythm that unites its people, its architecture, and its beliefs. The soul of Africa's civilization is **life itself**. It flows through the continent's rivers, it speaks through its forests, and it sings through its people.

GERD+ embodies that soul in motion. It transforms a river into light, showing that human creation can enhance, not exhaust, nature. The Pan-African Canal extends that spirit, linking nations the way veins link organs. Energy, water, and ideas move in harmony, keeping the continental body alive and balanced.

This civilization does not see nature as background; it sees it as family. Its power plants are gardens, its cities are living forests, and its economy is a circle, not a line.

From Empire to Ecology

Africa's living civilization will not resemble the empires of the past - built on conquest and competition. Its strength will not come from domination, but from **ecological intelligence** - the ability to design prosperity that restores rather than destroys.

Power in this civilization will not be measured by control over land but by care for it. Leaders will not expand borders; they will expand harmony. Progress will not be counted in tons of consumption, but in hectares of renewal.

It will be the first global civilization to turn away from extraction and toward **coexistence**.

The Three Pillars: Energy, Culture, and Balance

Africa's living civilization will rest on three interconnected pillars:

1. **Energy as Life** – Clean power from GERD+, solar networks, and hydrogen systems will circulate like blood, sustaining cities, villages, and industries. Every home will generate or store part of

the energy it uses, ensuring that abundance is shared, not hoarded.

2. **Culture as Connection** – Music, language, and art will guide how technology is used. Culture will not lag behind innovation; it will **lead** it, reminding progress to remain human.
3. **Balance as Law** – Every policy, every invention, every design will follow the principle of regeneration: what is taken must be returned, what is changed must be renewed.

These three pillars will make Africa's civilization not only sustainable but **symbiotic** - thriving because everything it touches continues to live.

The Living Architecture

In this new civilization, buildings will breathe. Cities will recycle their own air and water. Construction will draw from nature's wisdom - strong yet flexible, efficient yet beautiful. Materials will come from local, renewable sources; waste will return to the cycle as new value.

Skyscrapers will resemble vertical gardens. Housing developments will merge with forests. The Pan-African Canal will host floating solar farms and green ports - modern marvels shaped by ecological humility.

It will be architecture not of dominance, but of **dialogue** with the earth.

The Moral Character of Civilization

Africa's living civilization will also redefine ethics. It will see humanity not as master of the planet but as guardian. Progress will be measured by how gently we live, not how quickly we advance.

In this moral framework, every invention must answer one question: *Does it preserve life?* If it does, it is progress. If it does not, it is distraction.

This principle - simple yet profound - will guide governance, trade, and education. It will ensure that science serves compassion and that wealth serves wisdom.

Knowledge as Continuity

African civilization will not begin anew; it will **continue** the story of life that started here. The ancient understanding of harmony with the environment, once forgotten, will return in modern form - coded into policies, technologies, and cultural norms.

GERD+ and the Pan-African Canal are modern expressions of ancient truth: that flow is the essence of life. When systems flow freely - energy, water, culture, and cooperation - civilization remains young, no matter its age.

Unity Through Living Systems

The living civilization will erase the illusion of division. Nations will trade electricity as easily as words, and borders will become corridors of collaboration. Africa will no longer be seen as a collection of fifty-four states, but as **one living network** - each part unique, yet inseparable from the whole.

Unity will not come from treaties alone, but from systems that **depend on each other**. The grid, the canal, and the forests will make separation illogical, and cooperation natural.

A Civilization That Teaches the World

The ultimate purpose of Africa's living civilization is not to dominate the world, but to **guide it**. While others struggle to balance wealth and wellbeing, Africa will offer a model - a civilization that thrives by healing, that grows by giving, and that measures success by serenity.

The world will look to Africa not for resources, but for **reason**. It will study how Africa learned to live again, how it turned its pain into philosophy, and how it built a modern world that still felt human.

The Living Light of Civilization

In this civilization, light will not only illuminate space - it will illuminate spirit. Energy from GERD+ will power cities, but it will also power peace. The glow of hydrogen lamps and solar grids will symbolize more than progress; it will symbolize rebirth - the moment Africa remembered that civilization is not built with cement, but with compassion.

This is the Africa that awaits - not the Africa of survival or imitation, but the **Africa of life**.

A civilization that grows like a forest: rooted in its past, connected in its present, and endless in its future.

11.0 THE DAWN OF THE AFRICAN CENTURY

Every age of humanity has had its center of light. Once it was the valleys of the Nile. Later it was the rivers of Asia, the ports of Europe, and the engines of the West. Now, the light is moving south - returning to where it first began. The twenty-first century is not only a new chapter in history; it is **the Dawn of the African Century**.

This dawn does not arrive with noise. It comes with calm confidence - the rising glow of a continent that no longer waits for permission to shine. It comes from rivers like the **Blue Nile**, from systems like **GERD+**, from projects that bind rather than divide, and from people who have decided to turn their challenges into design.

The Century of Renewal

Africa's century will be defined not by conquest or extraction, but by **renewal**. Energy will renew itself through hydrogen and solar systems. Water will renew itself through canal networks and restored wetlands. Economies will renew themselves through circular industries.

This is the first century in human history when growth can mean healing - when progress and preservation become one act. The continent that suffered most from others' exploitation will lead the world into an age of balance.

A New Kind of Power

The power of the African Century is not domination; it is **integration**. Africa's strength will not come from the size of its armies or its wealth, but from the coherence of its systems - its ability to connect energy, knowledge, and compassion into a single flow.

GERD+ and the Pan-African Canal symbolize this power: unity through design. When nations share light and water, they share destiny.

The grids, pipelines, and forests of the African Renaissance will form the physical expression of peace - cooperation made tangible.

The Century of Intelligence

The African Century will be remembered as the time when the continent's greatest resource - its intelligence - was finally recognized and unleashed.

Its young people will not migrate in search of opportunity; they will create opportunity at home. Its researchers will not borrow ideas; they will originate them. Its engineers, architects, and farmers will not copy - they will **compose**.

Artificial intelligence will serve African intelligence - amplifying local knowledge instead of replacing it. Technology will speak the languages of the land, shaped by the logic of its rivers and seasons.

This will be the century when Africa stops being studied and starts being **understood**.

The Century of Balance

In a world still addicted to speed and consumption, Africa's greatest contribution will be balance. The continent will show that it is possible to grow without exhaustion, to industrialize without pollution, and to modernize without losing identity.

Forests will expand even as cities rise. Rivers will be protected even as they power turbines. Progress will move at the pace of peace - steady, thoughtful, and inclusive.

This will be the century when balance becomes the new measure of greatness.

The Century of Unity

Political borders will no longer define Africa's destiny; systems will. The flow of electricity, hydrogen, food, and ideas will knit the continent into one living fabric.

The **African Union** will evolve from aspiration to architecture - a governing network of shared systems, transparent markets, and cooperative leadership. Nations will remain sovereign but act symphonically - each instrument playing its part in the harmony of a continental design.

The Pan-African Canal will be the symbol of this unity: a blue thread of life connecting seas, economies, and hearts.

The Century of Dignity

Perhaps the most profound feature of the African Century will be dignity - the quiet pride of a people who no longer measure themselves through others' eyes.

Energy independence, food security, and educational excellence will restore that dignity in daily life. Every citizen will know that their labour, their rivers, and their sunlight feed a civilization that stands on its own.

The century will belong to a generation that has learned that dignity is not granted - it is grown.

The Century of Leadership

Africa's leadership will be moral before it is material. The continent's journey from exploitation to regeneration will give it the authority to lead by example.

When other regions face water scarcity, Africa will offer models of balance. When the world seeks clean fuel, Africa's hydrogen will supply it. When humanity searches for wisdom, Africa's philosophy of living systems will remind it how to breathe again.

This is leadership not through dominance, but through **direction** - leading humanity back to harmony.

The Century of Culture

The African Century will not only build infrastructure; it will build imagination. Its films, music, and literature will express a new identity - confident, creative, and conscious.

Art will no longer be a reflection of pain; it will be a declaration of possibility. The rhythms of Africa will become the heartbeat of the world again - not as entertainment, but as enlightenment.

The soul of the continent will return to its rightful place: the source of beauty that teaches humanity how to feel.

A Light for the World

The dawn of the African Century is not a rise against others but a rise **for all**. As Africa heals itself, it will heal the world - reducing emissions, restoring forests, balancing trade, and reminding civilization of humility.

Its light will not blind; it will guide. Its energy will not exploit; it will empower. Its success will not separate; it will connect.

The world will look toward Africa not as a continent of need, but as a **continent of meaning** - a living example of what civilization can become when it learns to grow like nature: patiently, purposefully, and peacefully.

The First Light

Every dawn begins with a single spark. That spark has already been lit - in the turbines of GERD+, in the flow of the Pan-African Canal, in the dreams of millions who now believe that change is not only possible but natural.

The light of this century is African because it rises from life itself. It is the same light that began civilization thousands of years ago - now returning home, brighter and wiser.

This is not the century Africa joins history. It is the century **history joins Africa**.

11.1 The Age of African Imagination

Every great age begins in the mind before it is built by the hand. The **Age of African Imagination** is that new beginning - a time when Africa no longer imitates the dreams of others, but imagines its own. It is the age when vision becomes the most valuable resource, when creativity becomes the new energy, and when hope itself becomes a form of power.

The imagination of Africa is vast because it was never fully used. It waited patiently - buried under years of limitation, interrupted by colonization, and silenced by dependency. But imagination cannot die; it can only sleep. And now, it is waking - bright, confident, and ready to reshape the world.

From Recovery to Creation

The last century was Africa's age of recovery - rebuilding, stabilizing, surviving. The century now unfolding is the age of **creation**. The continent no longer defines itself by what it lost but by what it can build.

GERD+, the Pan-African Canal, and the living circular systems of energy and ecology are examples of imagination turned into structure. They are not copies of global models; they are **African inventions**, drawn from Africa's geography, culture, and philosophy.

The same creative spirit will extend into every field - from architecture that grows like forests to economies that recycle like rivers, from schools that teach in native languages to art that speaks across borders.

The African imagination is not abstract; it is **applied poetry** - turning ideas into living systems.

The Courage to Dream

Imagination requires courage - the courage to dream beyond fear, beyond approval, beyond imitation. For too long, Africans were told to be realistic, to fit into systems built elsewhere. But the future belongs to those who can see what others cannot.

This age will be shaped by leaders, artists, scientists, and farmers who dare to think differently - who look at deserts and see farms, who look at rivers and see networks, who look at sunlight and see cities of light.

Courage will no longer mean fighting enemies; it will mean **believing in beauty**.

Education for Imagination

The new Africa will train imagination the way the old world trained obedience. Classrooms will become laboratories of creativity, not memorization. Students will learn to design, not just to follow instructions.

Science and art will reunite - because invention and inspiration are the same act. Engineers will study ecology; poets will study energy. Imagination will no longer be limited to the arts; it will become the foundation of all disciplines.

Every schoolchild will be encouraged to ask the most African question of all: *What can life become if we care for it?*

Imagination as Infrastructure

The greatest infrastructure Africa can build is not roads or dams but **ideas** - systems of thought that guide physical creation. GERD+ was built because someone imagined a river as more than a flow of water; they saw it as a flow of destiny.

In the same way, future projects will be born from imagination that sees connections others overlook - waste as energy, canals as communication lines, forests as factories of balance.

When imagination becomes institutionalized, innovation becomes inevitable.

Cultural Rebirth

The Age of African Imagination is also an age of **cultural renewal**. The continent's music, stories, and symbols will again become the tools of education, diplomacy, and design. Culture will not just reflect life; it will **direct** it.

Films will show not poverty but possibility. Literature will explore not survival but meaning. Fashion, architecture, and digital design will merge ancestral aesthetics with futuristic form - making Africa the creative compass of the world.

This cultural renaissance will remind humanity that beauty is not decoration; it is direction - the proof that civilization can still feel alive.

Technology Shaped by Imagination

Africa's technology will not be imported imitation; it will be adaptive creation. Renewable systems will draw from natural logic, just as digital systems will draw from human empathy.

Artificial intelligence developed in Africa will learn from indigenous philosophies - from the communal logic of Ubuntu, from the ecological rhythm of seasons, from the linguistic fluidity of African languages. It will not dominate humanity; it will **understand** it.

When technology is guided by imagination rooted in values, it becomes art in motion - science with soul.

The Imagination of Peace

Perhaps the highest form of imagination is the ability to see peace not as the absence of conflict but as the **presence of balance**. The Pan-African Canal and GERD+ are peace projects disguised as infrastructure - channels that carry understanding as much as water and energy.

The African imagination will prove that reconciliation can be engineered, that cooperation can be designed, that unity can be built like architecture - patiently, beautifully, and permanently.

Global Inspiration

The world will look to Africa for imagination the way it once looked to Greece for philosophy or to Europe for science. It will find in Africa the models it needs for survival - models of circular economies, regenerative agriculture, community-centered governance, and compassionate technology.

The African imagination will become humanity's mirror, showing that the future can be bright without being blinding, modern without being mechanical.

The Light of the Mind

In the end, imagination is light - invisible but powerful, intangible but necessary. The Age of African Imagination is the flowering of that light. It is the stage when Africa stops asking what it can learn from the world and begins teaching the world what it has always known: that the greatest source of power is the ability to **see life as possibility**.

The turbines of GERD+ spin because of water, but their true energy comes from the imagination that conceived them. The forests grow because of rain, but their pattern is shaped by the imagination of care. The Pan-African Canal flows because of engineering, but its purpose is carried by vision.

When imagination becomes the heart of governance, science, and culture, the continent becomes more than alive - it becomes **creative consciousness** made visible.

The Age of African Imagination is the age when the mind of a continent becomes its sunrise. It is the dawn not just of Africa's century, but of humanity's renewal - a reminder that the future is not something to predict, but something to **imagine wisely and build together**.

11.2 The Children of Light

Every civilization has a generation that carries its rebirth. For Africa, that generation is already here - the youth who refuse to inherit despair, who see light where others saw lack. They are **the Children of Light** - born into challenge, but destined for creation; raised in struggle, but driven by purpose.

They are not only the future of Africa; they are its proof that the continent's renewal is already alive. Each one of them is a spark - and together, they are the dawn that no night can silence.

Born in Transition

The Children of Light are growing up in a time of transformation. They belong to neither the old world of scarcity nor the chaotic world of excess. They stand at the meeting point - between darkness and dawn, between dependence and dignity.

They inherit both the pain of the past and the promise of the future. But instead of being crushed by that weight, they are learning to **turn it into strength**. They are not victims of history; they are designers of tomorrow.

Their lives bridge worlds - rural and urban, traditional and digital, local and global. And from that bridge will rise a new kind of wisdom: the ability to connect everything without losing the self.

Raised by Light

The Children of Light are the first generation to grow under the power of clean energy. Their schools are lit by solar and hydropower; their homes are powered by systems like **GERD+** and the **Pan-African Canal**. They will never know a world where darkness was normal, because light - both literal and symbolic - will define their reality.

Electricity gives them access to learning, technology, and connection. But more than that, it gives them a sense of possibility. They see that their rivers, their sun, and their forests are not burdens but blessings.

They are the generation that will grow with the belief that energy can be infinite - because creativity is infinite.

Educated in Renewal

The education of the Children of Light will be different. They will not be trained only to find jobs, but to **create systems**. They will learn how power flows, how water breathes, how forests regenerate, and how technology can serve life.

Their classrooms will extend into fields, laboratories, and communities. Learning will not end with graduation; it will continue through creation.

In their schools, teachers will not only teach subjects - they will teach **purpose**. The African curriculum of the future will not begin with what the world already knows, but with what Africa must now imagine.

The Spirit of Responsibility

The Children of Light will grow up knowing that light is both a gift and a responsibility. It must not be wasted, and it must not be used to harm. The same rivers that give them power also feed the land. The same forests that absorb carbon also give breath to the world.

Their leadership will be founded not on privilege, but on **guardianship** - the understanding that to have power is to protect life.

They will lead not through competition, but through cooperation. They will measure success not in wealth, but in wellbeing. And they will see progress not as consumption, but as contribution.

Unity as Strength

The Children of Light will not see themselves as citizens of isolated nations but as part of a **living continent**. They will collaborate across borders, languages, and fields - connected by shared energy, shared technology, and shared vision.

The Pan-African Canal and continental energy grid will be their inheritance - but also their responsibility. They will maintain and expand it, not as a political project, but as a moral covenant between generations.

In them, the dream of unity will stop being a slogan and become a **daily reality**.

Creativity as Identity

For these children, creativity will not be a talent; it will be a **way of thinking**. They will see problems as opportunities, and scarcity as a challenge to invent.

From their villages will come engineers who design microgrids; from their cities will come artists who tell stories of restoration; from their universities will come thinkers who merge science with soul.

Their imagination will no longer be a side of culture - it will be the **core of civilization**.

Technology with Humanity

The Children of Light will live in a digital Africa, but their technology will be guided by empathy. They will use machines not to replace people, but to reconnect them. Artificial intelligence will serve human understanding, not replace it.

Technology will not isolate; it will **illuminate** - helping farmers predict rainfall, students access global classrooms, and communities manage shared resources transparently.

In their hands, progress will regain its original meaning: *to move forward without losing what matters.*

The Return of Belief

Perhaps the greatest change in this generation will be spiritual. The Children of Light will believe again - in their continent, in themselves, in the future.

They will not seek validation from the outside world; they will create new standards from within. They will not dream of leaving Africa; they will dream of **lifting** it.

Their faith will not be in miracles, but in **motion** - the quiet miracle of doing things right, again and again, until light becomes normal.

The Generation of Balance

They will be the first generation to live in harmony with their environment - not because they are told to, but because it is natural to them. Clean air, green power, and circular systems will not be political debates but cultural instincts.

They will see balance not as a target, but as a way of life. In them, the ancient African wisdom of coexistence will meet the modern science of sustainability.

The Keepers of the Flame

The Children of Light are the keepers of a flame older than history - the same flame that once guided humanity's first steps, now reborn as hydrogen light, solar brilliance, and inner awareness.

They are the bridge between memory and possibility, between what Africa was and what it is becoming.

And when they look to the horizon, they will not ask what the future holds - they will ask **what the future needs**, and they will build it.

The Dawn in Their Eyes

The Children of Light will walk with calm confidence - not arrogant, not uncertain. They will know that the light in their hands comes from the same sun that burns above every nation, but their task is unique: to remind the world that progress must serve life.

They will not be defined by their parents' struggles, nor by others' expectations. They will define themselves by their contribution to the circle of renewal.

Through them, Africa's dawn will no longer be a hope - it will be a **human reality**.

11.3 The Light Beyond the Continent

When a light grows strong enough, it no longer belongs to one place. It travels. It reflects. It becomes a mirror through which others see themselves. The light that rises from Africa - the light of renewal, balance, and wisdom - is destined to reach far beyond its borders. It will not conquer the world; it will **complete** it.

The **Light Beyond the Continent** is the natural expansion of Africa's living civilization. What begins with GERD+, the Pan-African Canal, and the circular systems of energy will become a model for humanity - a practical and philosophical gift to a world that has forgotten how to live gently with its own power.

A Light That Teaches

Africa's light will not spread through speeches or sanctions, but through **example**. Nations will come to learn how the continent powers itself without pollution, irrigates without depletion, and grows without waste.

They will study how hydrogen fuels ships that leave no smoke behind, how forests expand even as cities rise, and how rivers are managed through cooperation rather than competition.

This is how Africa will teach - not by preaching, but by **proving** that harmony is possible.

The GERD+ model will be replicated around the world, showing that large-scale hydropower, when guided by wisdom, can heal rather than harm. The Pan-African Canal will inspire new systems of shared infrastructure that link regions through peace, not profit.

Africa's projects will become the new textbooks of sustainability - written not in theory, but in soil, sunlight, and success.

A New Global Ethic

The light beyond the continent is also **ethical**. Africa will remind humanity that progress must be measured not by speed or wealth, but by the wellbeing of the Earth. It will challenge the world to replace extraction with exchange, and dominance with dialogue.

This moral clarity will be Africa's greatest export. Its message will be simple and profound: *grow, but do not wound; build, but do not break; advance, but do not forget.*

The world, weary of endless crises, will find in Africa a calm center - a place that demonstrates how to rise without rivalry.

The Energy Bridge

Africa's clean energy will flow beyond its shores. Hydrogen from GERD+ and the canal corridors will power international ships, planes, and industries. The continent will supply not only energy but **stability** - dependable, renewable, and moral power.

But Africa will not export blindly. Its agreements will reflect fairness and reciprocity. Energy partnerships will include technology transfer, education exchange, and environmental accountability. The light will not be given away - it will be **shared intelligently**.

In this way, Africa's rise will also raise others - creating a global flow of energy guided by justice.

Cultural Light

The cultural light of Africa will travel even faster than its hydrogen. Music, art, design, and literature will shape global imagination. The world will look to Africa not for rhythm alone, but for **meaning** - for stories that restore humanity's connection to life.

The continent's artists will speak of regeneration, its architects will design with humility, and its thinkers will redefine what civilization means in an age of climate awareness.

African wisdom will re-enter the global vocabulary: the principle of Ubuntu, the respect for land, the idea that "I am because we are."

In a time when technology risks isolating humanity, Africa's cultural light will reconnect it.

The Light of Cooperation

Africa's systems will not stop at the edge of the continent. Its canal routes, energy grids, and digital corridors will link to Asia, Europe, and South America - forming new global networks based on renewable trade and shared resilience.

The **Pan-African Canal**, when extended to global ports, will become a highway of peace, carrying clean fuel and clean ideas. The energy that begins in African rivers will light cities across oceans, proving that interdependence can be noble when guided by equality.

The world will no longer see Africa as a receiver of systems but as a **giver of flow** - a source of motion, not memory.

Diplomacy of Light

Africa's diplomacy will be defined not by power politics, but by **planetary purpose**. The continent will lead climate discussions not as a victim seeking aid, but as a teacher offering wisdom.

Its negotiations will focus on shared survival, not short-term gain. Its policies will encourage fairness in carbon exchange, green finance, and technology sharing.

Africa's voice will become the moral compass of the international community - calm, reasoned, and rooted in lived experience.

The Human Light

Beyond technology and diplomacy, Africa's greatest gift to the world will be human. Its philosophy of coexistence, its patience, its emphasis on community - these will help heal the divisions of a restless planet.

The continent will remind humanity that progress without empathy leads to emptiness. It will show that wealth without warmth is failure, and that knowledge without kindness is blindness.

In this way, Africa's light will not simply illuminate; it will **humanize**.

The Circle Expands

As Africa's systems and values spread outward, they will form a **circle of continents** - a planetary version of the same circular philosophy that governs its own development. Each region will contribute to the cycle of renewal, balancing resources and responsibilities.

The world will begin to operate more like a forest than a factory - diverse, interlinked, and self-sustaining.

A Light Without Borders

Ultimately, the light beyond the continent will erase the old map of privilege and prejudice. It will remind humanity that light has no nationality, no colour, and no hierarchy.

The African Century will therefore not end in Africa; it will flow through it - a gift to every place willing to learn how to live.

The world will one day realize that the light it follows came not from invention, but from **rediscovery** - the rediscovery of life's simplest truth: that everything thrives when it shares.

And when that day comes, the Earth itself will glow a little softer, a little wiser - because Africa, the first home of humanity, will have become its **final teacher**.

11.4 The Final Sunrise: Africa's Gift to Humanity

Every civilization has a moment when it gives back to the world what the world once took from it. For Africa, that moment is near - a sunrise that will not fade, a light that carries both memory and mercy. The **Final Sunrise** is not the end of a story; it is the beginning of understanding. It is the hour when Africa's quiet rise becomes humanity's shared awakening.

Africa's gift is not gold, oil, or minerals. It is wisdom - the wisdom of balance, the patience of rivers, the humility of soil, and the generosity of light. These are not commodities; they are truths. They remind humanity that progress is sacred only when it sustains life.

The Meaning of the Final Sunrise

The word *final* does not mean last. It means **fulfilled** - a completion of a long circle of history. The light that once left Africa millennia ago, carried by the first humans, now returns home. It carries knowledge, technology, and experience - but it also returns with the same question: *Can humanity live without wounding what gives it life?*

Africa's answer is yes. GERD+, the Pan-African Canal, and the circular systems that connect them are that answer made visible - designs that prove civilization can advance without destruction, grow without greed, and build without breaking the balance of nature.

This sunrise will not blind; it will bless. It will light the world gently, as dawn lights the mountains after a long night.

A Civilization That Completes Itself

Africa's journey is not one of revenge or replacement. It is a **completion** - a civilization closing the cycle it began. The continent that once taught humanity how to walk, grow food, and build shelter will now teach it how to live in harmony with its planet.

Where other civilizations sought to control nature, Africa's living civilization will show how to **coexist** with it. Where others divided progress from peace, Africa will merge them. Where others separated spirit from science, Africa will weave them together.

This unity of mind and matter, of system and soul, is Africa's most profound contribution to humanity's future.

The Return of the Human Spirit

Africa's gift to the modern world is not only environmental; it is spiritual. It offers a way for humanity to rediscover its wholeness.

The forests teach patience. The rivers teach cooperation. The sunlight teaches equality. The continent's living systems remind the world that knowledge is not enough without care.

Through its philosophy of living balance, Africa will restore the **human spirit** - weary from centuries of division and competition. It will reintroduce harmony where ambition once reigned, calm where speed once ruled, and community where isolation once spread.

This is Africa's silent revolution: a revolution of **returning to what matters**.

The Language of Light

As the African Century unfolds, the continent's voice will change the world's language. Words like *growth*, *progress*, and *development* will gain new meanings. Growth will mean regeneration; progress will mean balance; development will mean wisdom.

Light itself will become a metaphor for **understanding**, not consumption. To enlighten will mean to heal.

This new language will travel through art, policy, technology, and culture - guiding humanity toward a gentler future.

The Gift of Renewal

Africa's final gift is **renewal** - not just of power and nature, but of purpose. It reminds the world that nothing is too broken to be reborn, that deserts can bloom again, and that people can rediscover faith in one another.

The hydrogen systems of GERD+, the green corridors of the Pan-African Canal, the forests of the living continent - all show that regeneration is not only natural but inevitable when guided by care.

This philosophy will ripple outward - influencing global systems of energy, agriculture, and economics. Humanity will learn that every crisis is simply the earth asking us to begin again.

The Quiet Example

Africa will not force its light upon the world. It will simply **become** what it teaches. The continent's clean energy systems, circular industries, and ethical models of governance will speak louder than any declaration.

As other regions face droughts, depletion, and division, they will look south and see something different - a civilization thriving without haste, calm in its confidence, and radiant in its renewal.

Africa's success will not demand attention; it will **attract** it - as truth always does.

A World That Learns to Breathe Again

Humanity's survival will depend on rediscovering what Africa never forgot: that the Earth is alive. When the rest of the world remembers this, inspired by Africa's example, a new era will begin - not of nations, but of balance.

Rivers will be managed with care; energy will be generated with humility; food will be grown with gratitude. Humanity will breathe again - slowly, deeply, wisely.

The planet will not be saved by technology alone, but by **the intelligence of care** - an intelligence Africa has carried since the beginning of time.

The Circle Completed

The Final Sunrise will mark the closing of a great circle. Humanity's story began in Africa, wandered through millennia of ambition, and now returns to its source to learn simplicity once more.

This is not a step backward; it is a **return forward** - the merging of ancient wisdom with modern knowledge, the balance between earth and mind.

When that circle closes, civilization will become whole again.

Africa's Eternal Gift

The light of this sunrise will never fade, because it is not made of electricity or fire - it is made of **understanding**. It lives in the realization that the future belongs not to the strongest or the richest, but to those who live in harmony with the flow of life.

That is Africa's gift to humanity: not power, not wealth, not dominance - but **balance**. The wisdom to grow without greed, to lead without pride, and to shine without burning.

When the world accepts this gift, a new dawn will begin - not just for Africa, but for all.

The Final Sunrise will not end a story. It will begin the story of a planet finally at peace with itself - a world lit not by ambition, but by awareness.

And when that light spreads across the seas, soft and steady, humanity will remember what Africa always knew: that **the greatest power is not to take light, but to become it**.

EPILOGUE

Africa's rivers carry more than water; they carry possibility. From the strength of GERD+ and the flow of the Pan-African Canal emerges a new chapter - an Africa that generates, stores, and shares **living energy**.

Hydrogen becomes the heartbeat of this transformation - pure, clean, and endlessly renewable. And from hydrogen comes **ammonia** - the quiet companion that carries this energy across oceans and into the soil.

Ammonia is not a chemical of industry alone. It is the **bridge between energy and life**. It stores hydrogen safely, travels easily, and returns to the land as nourishment. It connects power generation with agriculture, and technology with nature.

In this cycle, Africa finds its rhythm. Electricity from GERD+ produces hydrogen; hydrogen forms ammonia; ammonia feeds crops and exports clean energy. Nothing is wasted - every process feeds another. This is the meaning of **systems thinking** - seeing energy, food, water, and life as one continuous design.

The future Africa envisions is not one of dependence, but of design. An Africa that exports hydrogen, ammonia, and fertilizers made from its own sunlight and rivers. An Africa that feeds itself while powering others. An Africa where technology grows like nature - in circles of renewal.

The world's next energy hub is not defined by oil fields or gas terminals; it is defined by rivers, light, and intelligence.

It is **GERD+ - Africa's Future Energy Hub**.

This is not a conclusion. It is a beginning - a living flow of clean energy, green soil, and awakened minds. For the young Africans who will inherit this future: your generation will not burn the Earth to survive; you will **illuminate it to live**.

AUTHOR'S REFLECTIONS

This book was not written to describe a project; it was written to describe a **possibility**. The ideas of **GERD+** and the **Pan-African Canal** came from long years of reflection - from looking at Africa's rivers, sunlight, and silence, and asking a simple question: *What if we began to live in harmony with what we already have?*

I did not imagine these systems as technical blueprints, but as **symbols of consciousness** - signs of how Africa could think again as a living whole. Every page of this book was guided by that belief: that the future of this continent does not depend on imitation, but on **imagination**.

Africa's greatest wealth has never been its resources, but its rhythm - the natural intelligence of balance that flows through its land and people. When we design our systems according to that rhythm, we rediscover what was never lost: the ability to grow without destruction, to rise without rivalry, to shine without arrogance.

GERD+ became, for me, a mirror of that idea. It represents how a single river, if understood deeply, can generate more than power - it can generate peace. It shows how water can unite where politics divide, and how light can heal where history once hurt.

The **Pan-African Canal** followed as its companion vision - a line not of separation, but of connection. It carries more than water; it carries hope. It connects deserts with gardens, highlands with coasts, nations with purpose. Together, GERD+ and the Canal symbolize the awakening of a continent that has remembered its calling: not to compete, but to **complete** - to bring wholeness back to a fragmented world.

While writing, I found myself returning to one truth: Africa does not need to begin from scarcity. It needs to begin from **self-awareness**. Everything essential is already here - the energy of the sun, the flow of the rivers, the resilience of its people, and the wisdom of its soil. What remains is the courage to trust that abundance, to organize it, and to share it.

This book is therefore not a conclusion, but a **continuation** - a reminder that the circle of life and light never ends. Each idea, each dream, each act of care adds to that circle. And if enough of us choose to live by it, Africa will not only transform itself; it will quietly guide the world toward balance.

I believe in that Africa - an Africa that lives rather than survives, that designs rather than depends, that teaches rather than waits. An Africa that builds systems that breathe, economies that restore, and cities that remember the silence of the soil beneath them.

This belief is not a hope; it is a certainty born from experience and observation. I have seen light return to the places that were once dark - not just through electricity, but through **understanding**. When knowledge and kindness work together, light becomes permanent.

As I close this work, I do so not with an ending, but with gratitude - gratitude to every African who believes that renewal is possible, to every young mind that imagines a different tomorrow, and to the rivers and skies that have carried our stories since the beginning.

Africa's dawn has already begun. It will not arrive all at once; it will rise gently, as all true light does - steady, honest, and full of meaning.

And when that light fills the horizon, it will remind us that the greatest power of all is not in conquering the world, but in **living beautifully within it**.

Lakew Gebrekirstos Buli
Addis Ababa, Ethiopia
2025

BIBLIOGRAPHY

African Union. (2020). *Agenda 2063: The Africa we want.* Addis Ababa: African Union Commission.

Aircrete Europe. (2023). *Autoclaved aerated concrete: Innovation and sustainability in construction.* The Netherlands: Aircrete Europe Publications.

Asafu-Adjaye, J. (2018). *Energy, environment, and economic transformation in Africa.* London: Routledge.

Devarajan, S., & Monga, C. (2019). *The economics of Africa's energy transformation.* Washington, DC: World Bank.

Elkington, J. (2018). *Green swans: The coming boom in regenerative capitalism.* London: Fast Company Press.

Hafner, M., & Tagliapietra, S. (2019). *The geopolitics of the global energy transition.* Cham: Springer.

Kammen, D. M., & Kirubi, C. (2020). *Renewable energy pathways in sub-Saharan Africa.* Nairobi: African Energy Policy Institute.

Moyo, D. (2018). *Edge of chaos: Why democracy is failing to deliver economic growth-and how to fix it.* London: Penguin Books.

Nkrumah, K. (1963). *Africa must unite.* London: Panaf Books.

Odhiambo, N. M. (2021). *Energy poverty and sustainable development in Africa.* Johannesburg: University of South Africa Press.

Owusu, P. A., & Asumadu-Sarkodie, S. (2016). A review of renewable energy sources, sustainability issues, and climate change mitigation. *Cogent Engineering, 3*(1), 1167990. https://doi.org/10.1080/23311916.2016.1167990

Rockström, J., & Klum, M. (2015). *Big world, small planet: Abundance within planetary boundaries.* New Haven, CT: Yale University Press.

Sachs, J. D. (2015). *The age of sustainable development.* New York: Columbia University Press.

Schumacher, E. F. (1973). *Small is beautiful: Economics as if people mattered.* London: Blond & Briggs.

Sörgel, H. (1932). *Atlantropa: A plan to unite Africa and Europe.* Munich: Pilot Verlag.

United Nations Economic Commission for Africa (UNECA). (2022). *Renewable energy and green hydrogen in Africa: Policy frameworks and pathways.* Addis Ababa: UNECA Press.

World Bank. (2023). *Energy for development: Africa's pathways to a sustainable future.* Washington, DC: World Bank Publications.

Wright, J., & Tipping, D. (2019). *Hydrogen futures: Clean energy for the global economy.* Oxford: Oxford Energy Studies Institute.

CONCEPTS AND THEMES

GERD+ Africa's: Future Energy Hub

The ideas explored in GERD+: Africa's Energy Renaissance are not presented as technical prescriptions or political programs, but as **living concepts** - philosophical frameworks designed to inspire new ways of thinking about how Africa can grow, create, and coexist. The following concepts and themes summarize the core messages that run through the book, forming the foundation of a new African narrative built on balance, renewal, and responsibility.

GERD+ - The Philosophy of Living Energy

GERD+ symbolizes more than hydropower. It represents a new consciousness of energy - one that transforms natural flow into ethical power. The "plus" stands for expansion beyond engineering into ecology, economy, and empathy. It demonstrates how energy can regenerate rather than deplete, how water can unite rather than divide, and how progress can be measured not in megawatts alone, but in **meaning**.

The Pan-African Canal - Connection as Civilization

The Pan-African Canal is envisioned as a continental artery of life. It connects not only rivers and seas, but also economies, cultures, and climates. It represents the idea that **infrastructure can be diplomacy**, that water can become peace, and that unity can be designed. Through the Canal, Africa's geography transforms from a barrier into a bridge - a living symbol of cooperation and shared destiny.

The Living Continent

Africa is presented not as a resource, but as a **living system** - with its own intelligence, rhythm, and memory. Every river, forest, and community functions as part of an ecological body. When one part thrives, all benefit; when one suffers, all weaken. The concept of the Living Continent calls for a civilization modelled after nature itself: circular, adaptive, and whole.

Energy as Balance, Not Extraction

The book redefines energy as a relationship rather than a resource. True energy does not exploit; it circulates. GERD+, solar fields, and hydrogen systems together form a model of **regenerative power** - energy that flows like water, supporting life as it moves. This vision moves Africa beyond the extractive habits of the past toward a future where production and preservation become one continuous act.

Circular Africa - The Economy of Regeneration

Circular Africa is not only a strategy for waste reduction; it is a moral and ecological worldview. It envisions industries that return what they take, cities that recycle what they use, and societies that measure prosperity by how much they restore. In this theme, **the end of one process becomes the beginning of another**, ensuring that growth always leads to renewal.

The Carob and the Moringa - Living Symbols of Energy

These two trees embody the union of ecology and economy. The **Carob** teaches patience, endurance, and stability - the power of slow growth that lasts. The **Moringa** represents abundance and vitality - life that nourishes everything around it. Together, they form natural batteries of balance, illustrating how Africa's reforestation can be both an environmental and economic revolution.

The Circle of Life and Light

Throughout the book, the circle recurs as a central image - symbolizing wholeness, continuity, and harmony. In Africa's future civilization, nothing is wasted; everything returns. Light feeds life, and life feeds light. This circular rhythm is both scientific and spiritual - a design principle for how energy, water, and wisdom can sustain one another infinitely.

The Renaissance of Light

This theme marks Africa's emergence from centuries of limitation into an era of illumination. It is not just the spread of electricity, but the awakening of **awareness** - light as knowledge, as equality, and as peace. The Renaissance of Light celebrates the moment when Africa begins to power itself with its own resources and imagination, turning energy into identity.

The Africa of Systems and Wisdom

This concept describes the shift from isolated projects to interconnected systems. In the Africa of Systems and Wisdom, every structure - a dam, a canal, a forest - serves multiple purposes and functions in cooperation with others. This systemic design reflects the intelligence of nature and the ethics of balance. Wisdom replaces haste; harmony replaces competition.

The Eternal Flow of African Energy

Energy in this vision is continuous - it circulates between water, air, soil, and spirit. It does not end when it is used; it begins again. The Eternal Flow expresses Africa's identity as a **continent of movement and renewal**, where the same forces that once sustained life now sustain civilization. This theme teaches that power must always return to its source - purified and multiplied through purpose.

The Children of Light

The youth of Africa represent the continent's most powerful current of transformation. They are born into a world defined not by scarcity but by possibility. The Children of Light symbolize education, creativity, and moral courage - a generation that will not only use energy but understand it, not only consume but **create**. They are the living embodiment of the light that never ends.

Africa's Final Gift - Balance as Civilization

In its closing vision, the book presents Africa's greatest gift to humanity: **balance**. The ability to live with the Earth, not above it; to grow with compassion, not conquest. This is the true meaning of Africa's energy renaissance - not a rush toward power, but a return to peace. Through GERD+, the Canal, and the living philosophy they represent, Africa offers the world a path toward equilibrium - the realization that the highest form of progress is harmony.

All the concepts of GERD+: Africa's Energy Renaissance converge toward one truth: that the future belongs to those who understand ***life as a system of balance****. The Africa described in these pages is not an idea to admire - it is an invitation to act, to think, and to live differently.*

It is a continent that generates light without darkness, energy without waste, and strength without violence - a continent that remembers itself as a living whole.

ABOUT THE AUTHOR

Dr. Lakew Gebrekirstos Buli is the Founder and Chief Executive Officer of **Fusion Global Solutions Plc**, based in Addis Ababa, Ethiopia. He is an **African systems thinker and visionary author** whose work explores how Africa can rise through balance, integration, and self-reliance.

His approach **incorporates diverse theories of strategy, leadership, portfolio management, program management, project management, and systems thinking**, translated into an African context of renewal and cooperation. Through his *Living Africa Series*, Dr. Lakew presents a coherent philosophy of transformation - connecting energy, water, industry, and people into one living system capable of sustaining itself.

He writes and works with a single conviction: that Africa's future will not depend on imitation, but on **understanding and creation**. His thinking unites practicality with wisdom, vision with structure, and science with humanity.

Above all, **he sees himself as a human being** - guided by simplicity, humility, and a deep respect for life. His purpose is not only to build projects or write books, but to **help Africa remember its own intelligence** and to inspire a generation that designs systems that live.

ABOUT THE LIVING AFRICA SERIES

The Living Africa Series is a ten-book journey that reimagines Africa as a living, self-sustaining system - from strategic thinking to leadership, from clean energy to urban design, from water systems to circular industries, and finally to a continental philosophy of balance.

Together, the ten books form one continuous idea: that Africa can transform itself through wisdom, integration, and self-reliance - not by copying others, but by remembering its own intelligence and designing systems that live.

Book 1 – Strategic Africa: Shaping the Continent's Path to Purpose and Power

Defines Africa's strategic awakening - how a clear continental vision can turn resources into renewal, unity into strength, and ideas into collective direction. It lays the foundation for the philosophy of living systems that continues throughout the series.

Book 2 – Leadership for a Living Continent: Guiding Africa with Integrity and Wisdom

Reimagines leadership as service, consciousness, and balance. It calls for a generation of ethical African leaders who unite technical skill with humanity, guiding transformation through values rather than control.

Book 3 – Managing Africa's Project Portfolio: From Vision to Completion

Explains how Africa can plan, prioritize, and execute major programs that move ideas into measurable results. It shows how governance, management, and vision can work as one system - turning intention into impact.

Book 4 – GERD+: Africa's Future Energy Hub

Presents the GERD+ concept - combining hydropower, floating solar, and green hydrogen into an integrated African energy model. It demonstrates how clean energy can power industry, logistics, and hydrogen production, transforming Africa's role in the global energy map.

Book 5 – The Pan-African Canal: Africa's Waterway of Unity and Transformation

Envisions a transcontinental canal connecting landlocked nations to the sea - symbolizing the flow of African unity. It integrates water, energy, and trade corridors into one continental infrastructure of connection and peace.

Book 6 – Circular Africa: Building a Regenerative Continent

Explores circular economies and regenerative industries as the path from extraction to renewal. It shows how waste can become value, cities can breathe again, and Africa can lead the world in sustainable design.

Book 7 – Skyscraper Africa: Building Vertical Cities for a Rising Continent

Proposes a new model of African urban growth - vertical, efficient, and human. It connects housing, materials, and technology to create cities that rise with dignity, preserve land, and express African identity.

Book 8 – Electrify Africa: Powering a Living Continent

Examines how clean power - from water, sun, and hydrogen - can illuminate every home, factory, and mind. It reframes electricity as a force of dignity, productivity, and independence, essential for Africa's next leap forward.

Book 9 – The Vision: A Living Blueprint for Africa's Next Century

Calls for the rebirth of African imagination. It reminds readers that development begins with how we see ourselves, urging Africa to design its own future with confidence, creativity, and self-awareness.

Book 10 – The African Living System: Designing a Continent That Lives

Brings all ideas together into one living philosophy - Africa as an integrated organism of rivers, forests, people, and dreams. It defines the African Living System as a framework of renewal, harmony, and human purpose.

Together, these ten books form one living conversation - a map for Africa's awakening, guided by wisdom, responsibility, and light.

www.ingramcontent.com/pod-product-compliance
Lightning Source LLC
LaVergne TN
LVHW100318210926
848480LV00003B/1176